Rank Correlation Methods

FIFTH EDITION

Maurice Kendall
and
Jean Dickinson Gibbons

A CHARLES GRIFFIN TITLE

OXFORD UNIVERSITY PRESS
New York

Published in the USA by
Oxford University Press
200 Madison Avenue, New York, NY 10016
Fifth edition 1990

First published in Great Britain 1948

Library of Congress Cataloguing in Publication data
90-052649
Full library data is available upon request.

Printed in Great Britain

Preface to the Fifth Edition

My revision of Kendall's classic book has retained his original material as well as his style of writing. The previous editions of this book constitute an established institution which needed primarily to be made more reflective of the statistical literature and conventions of today.

The primary change in this edition is an attempt to update the references by expanding the notes and references sections at the end of each chapter. Unfortunately, the literature on the techniques covered by this book is now so voluminous that time demands made it impossible for an exhaustive search of the literature to be completed. Even if it had been possible, the exponential expansion of materials since Kendall's first edition in 1948 would make a complete listing not appropriate in a volume of this size. Every time I found a new reference, it led to several more. As a result, while the References have been expanded greatly, I make no claim that every relevant and important citation has been included. Therefore, I apologise to those authors whose contributions are not included here.

The secondary changes include more extensive tables of the null distribution of Kendall's tau (Table A.1) and Spearman's rho (Table A.2), plus a new table of the null distribution of Kendall's partial tau (Table A.11) in the Appendix.

Some additions and changes have been inserted in the text of Chapters 1, 3, 4, 6 and 8 in an attempt to clarify and update the exposition. Additional numerical examples have also been inserted in the text, and P-values are emphasized more than carrying out tests of significance. In addition, numerical problems have been added at the end of these same chapters so that readers can learn to apply the techniques explained in each chapter. All problems are based on real data concerning interesting research reported in the literature of many diverse fields. Chapter 13 has been rewritten. Appropriate problems for Chapters 2, 5, 7, 9, 10 and 12 are verification of the results given there.

Two sets of references are included at the end of the book. The first list, labeled References, gives citations for the articles and books referred to in the text, usually in the Notes and References section at the end of each chapter. These citations are from the statistical literature. The second list, labeled References for Applications, gives citations for the actual research situations that are described in the numerical examples and problems given in the text. These citations are to journals in various fields that use statistical methods to investigate research hypotheses.

I had originally planned to make substantial changes in notation but this proved to be imposible; it would only have led to more errors and inconsistencies than improvements. The end result is that the only changes in notation are the consistent use of t instead of τ for the Kendall sample tau coefficient; r_s instead of ρ_s for the Spearman sample rank correlation coefficient; Σ instead of S to indicate a summation; u and v respectively instead of t and u in the corrections for ties (this change was necessitated by the substitution of t for τ); A, B, C, etc. to denote the objects or members to be ranked; and X, Y, Z, etc. to denote the resultant rankings. I apologise for any notation changes that were missed in editing.

I hope that readers will write me about important omissions in references and inconsistencies in notation so that they can be incorporated in future printings.

I give sincere thanks to Professor Herbert A. David for encouraging me to undertake this project, to many colleagues who sent reprints and information about additional references, and to Phillip J. Ward for assistance in the literature search.

Finally, I acknowledge the summer support provided by the Research Grants Committee and the Manderson Graduate School of Business at the University of Alabama.

Jean D. Gibbons
Tuscaloosa, Alabama 35487
December 1988

Contents

Preface   iii

Chapter 1   The measurement of rank correlation   1
   Introductory remarks   1
   Kendall tau coefficient   3
   Tau as a coefficient of concordance   6
   Tau as a coefficient of disarray   7
   Spearman's rho   8
   Conjugate rankings   10
   Daniels' inequality   12
   The Durbin–Stuart inequality   12
   Stragglers   13
   Notes and references   20
   Problems   20

**Chapter 2   Introduction to the general theory of
rank correlation**   25
   The general correlation coefficient   25
   Tau as a particular case   25
   r_s as a particular case   26
   Product-moment correlation as a particular case   27
   Proof of Daniels' inequality   34
   Proof of the Durbin–Stuart inequality   34
   Spearman's footrule   37
   Notes and references   38

Chapter 3   Tied ranks   40
   Calculation of t 40
   Calculation of rho   43
   Application to ordered contingency tables   50
   Notes and references   54
   Problems   55

Chapter 4   Tests of significance   60
   The significance of t 62
   P-values   64

Continuity correction for S 65
Ties   66
The significance of r_s 69
Continuity correction for $\sum d^2$ 70
Tests in the non-null case   72
Applications to time-series data   76
Notes and references   78
Problems   79

Chapter 5   Proof of the results of Chapter 4   **91**
Exact distribution of t in the null case   91
Tendency of t to normality in the null case   93
Distribution of r_s in the null case   97
Joint distribution of t and r_s 99
Corrections for continuity   103
The non-null case   104
More exact treatment in the non-null case   109
r_s in the non-null case   113
Notes and references   115

Chapter 6   The problem of m ranking   **117**
The significance of W 121
Continuity correction for W 123
Estimation   124
Friedman test for randomised complete block designs   127
Incomplete rankings   129
Notes and references   133
Problems   135

Chapter 7   Proof of the results of Chapter 6   **144**
Notes and references   153

Chapter 8   Partial rank correlation   **154**
Notes and references   160
Problems   161

Chapter 9   Ranks and variate values   **164**
Concordances   164
Relation between ranks and variate values   165
Relation between t and parent correlation in the normal case   166
Relation between ρ_s and ρ in the normal case   169
Notes and references   171

Chapter 10   Proof of the results of Chapter 9   173
Correlation between ranks and variate values   173
Concordance   176
Notes and references   183

Chapter 11   Paired comparisons   184
Coefficient of agreement   188
Notes and references   194

Chapter 12   Proof of the results of Chapter 11   195
Notes and references   200

Chapter 13   Some further applications   201
Estimation of population consensus   201
Two group concordance   201
Comparison of n ranking with a criterion ranking   202
Uses of rank correlation in linear regression   202
Power and efficiency of rank correlation methods   203

Appendix Tables   204
1. Probability function of S and t (Kendall)   204
2. Probability of $\sum d^2$ (for r_s)   206
3. Probability function of the standard normal distribution   214
4. Random rankings of 20 (random permutations of the first 20
   natural numbers)   215
5. Probability function of S (for Kendall's coefficient of
   concordance)   224
6. Significance points of S (for Kendalls coefficient of
   concordance)   228
7. Significance points of Fisher's z distribution   229
8. Significance points of x^2 231
9. Probability function of d in paired comparisons   232
10. Probability function of Σ (for u)   234
11. Significance points of t_{XYZ} (for Kendall's partial rank correlation
    coefficients)   237

References   238
References for Application   256
Index   259

Chapter 1

The measurement of rank correlation

Introductory remarks

1.1 A cardinal number, such as three, is one that indicates quantity or size but not order except by comparison with another cardinal number. An ordinal number is one that indicates order or position in a series, like first, second, etc. When objects are arranged in order according to some quality which they all possess to a varying degree, they are said to be *ranked* with respect to that quality. The arrangement as a whole is called a *ranking*. The rank of each object indicates its respective position in the ranking.

1.2 It is customary, but not essential, to denote the ranks by ordinal numbers $1, 2, \ldots, n$, where n is the number of objects. Thus the object or individual which comes fifth in the ranking has the rank 5. In the sequel we shall often operate with these numbers as if they were the cardinal numbers of ordinary arithmetic, adding them, subtracting them and even multiplying them; and it is of some importance to realise exactly what such processes mean.

1.3 Suppose, for example, that an object has a rank 5 when the set of objects is ranked according to some quality X and a rank 8 according to a second quality Y. What is implied by saying that the difference of the ranks is 3? We cannot subtract 'fifth' from 'eighth'; but a meaning can be given to the process nevertheless. To say that the rank according to X is 5 is equivalent to saying that, in arranging according to X, four members are given priority over our particular member, or are *preferred* to it. Similarly, seven members are preferred in the ranking according to Y. Consequently the number of preferences in the Y-ranking exceeds the number in the X-ranking by 3; and this is not an ordinal number but a cardinal number, i.e. arises by counting.

This may strike the readers as a precious distinction which is hardly worth making at the present stage. If so, they can put it aside until it arises later. They should realise, however, from the outset that the numerical processes associated with ranking are essentially those of counting, not of measurement.

1.4 In practice, ranked material can arise in many different ways, some of which may be briefly mentioned:

(a) Purely as arrangement of objects which are being considered only by reference to their position in space or time. For instance, if we arrange a pack

1

of cards in some order and then shuffle them, the new order is a ranking which may be compared with the old to see whether the shuffling process is a thorough one. We are interested in the spatial arrangement alone—not, for example, in whether some objects are 'greater than' or 'less than' others in the intensity of a common quality.

(b) According to some quality which we cannot measure on any objective scale. For instance, we might rank a set of mineral specimens according to 'hardness' by some such simple criterion as saying that X is harder than Y if X scratches Y when the two are rubbed together. If X scratches Y and Y scratches Z then X will scratch Z, so that by making a number of comparisons we can rank the objects without ambiguity (unless two of them are equally hard, a special case we shall consider in Chapter 3). There is, however, no method of *measuring* hardness implicit in this approach. We can always decide whether X is harder than Y, but we cannot say that it is twice as hard without imposing some scale of measurement on the system.

(c) According to some measurable or countable quality. For instance, we may rank individuals according to height, or countries according to size of population. It may not always be necessary to carry out the actual measurements in such cases, as for instance, if we arrange a class of students in order of height 'by eye'; but the quality according to which the ranking is made is capable of practical measurement.

(d) According to some quality which we believe to be measurable but cannot measure for practical or theoretical reasons. For instance, we may rank a number of persons according to 'intelligence' on the assumption that there is such a quality and that individuals can be ranked according to the degree of intelligence which they possess. In Chapter 11 we shall consider a method which enables us to investigate in some cases whether these assumptions are legitimate. The reason for differentiating this case from that of paragraph (b) is that in the latter we know from physical considerations that ranking is possible, whereas in the former the possibility is a hypothesis.

1.5 In the theory of statistics a quantity which may vary from one member of a population to another is called a *variate*. In particular, a measurable quality provides a variate and, of course, a level of measurement that is interval or ratio scale. We can always rank a set of objects according to their position on a scale and may then replace variate-values by corresponding ranks. A ranking may then be regarded as a less accurate way of expressing the ordered relation of the objects—less accurate because it does not tell us how close the various objects may be on the scale of measurement. However, what the ranking loses in accuracy it gains in generality, for if we stretch the scale of measurement (and even if we stretch it differently in different regions) the ranking remains unaltered; in mathematical language it is *invariant* under stretching of the scale.

The ordinal level of measurement is lower than either the interval or ratio level because the arithmetic difference between ordinal measurements is not meaningful. If, for, example we rank from smallest to largest we know that the object with rank two is larger than the object with rank one but is not necessarily twice as large. Also, the difference between ranks one and two is not necessarily the same as the difference between ranks two and three, or eight and nine.

1.6 Historically the theory of ranks developed as an offshoot of the theory of variates. In the early stages ranks were regarded in the main as convenient substitutes for variate measurements to save time or trouble or to avoid the difficulties of setting up an objective scale. More recently they have been recognised as having an importance of their own, and in the ealier part of this book we shall consider ranking problems as such without any reliance on an underlying higher level of measurement. Our methods thus have very considerable generality. The relationship between ranks and variates will be discussed in Chapters 9 and 10.

Kendall tau coefficient

1.7 Suppose a number of boys are ranked according to their ability in mathematics and in music. Such a pair of rankings for ten boys, denoted by the letters A to J, might be as follows:

Boy:	A	B	C	D	E	F	G	H	I	J	
Math:	7	4	3	10	6	2	9	8	1	5	(1.1)
Music:	5	7	3	10	1	9	6	2	8	4	

We are interested in whether there is any relationship between ability in mathematics and music. A glance at these rankings shows that the agreement is far from perfect, but that some boys occupy the same or nearly the same position in both subjects. We can see the correspondence more easily if we re-arrange the math ranking in the natural order, as:

Boy:	I	F	C	B	J	E	A	H	G	D	
Math:	1	2	3	4	5	6	7	8	9	10	(1.2)
Music:	8	9	3	7	4	1	5	2	6	10	

What we wish to do is to measure the degree of correspondence between these two rankings, or to measure the intensity of *rank correlation*. We shall accordingly show how to construct a coefficient for this purpose which will be denoted by t and called the Kendall tau coefficient.

1.8 Such a coefficient should have the following three properties:

(a) if the agreement between the rankings is perfect, i.e. every individual has the same rank in both, t should be $+1$, indicating perfect positive correlation;

(b) if the disagreement is perfect, i.e. one ranking is the inverse of the other, t should be -1, indicating perfect negative correlation;

(c)   for other arrangements t should lie between these limiting values; and in some acceptable sense increasing values from -1 to 1 should correspond to increasing agreement between the ranks.

The first two of these requirements are only conventions, but are by far the most useful conventions to employ.

1.9   In the first ranking of (1.1) consider any pair of objects, say AB. Their ranks, 7, 4, occur in the inverse order (taking the natural order 1, ..., 10, as the direct order) and hence we will score for this pair the value -1. Had the pair been in the direct order we should have scored $+1$. In the second ranking the pair AB has ranks 5, 7, which is in the direct order, and we will, therefore, score $+1$ in this ranking.

We now multiply the scores for this pair in both rankings and hence arrive at the score -1. Evidently, for any pair, the resulting score is $+1$ if their ranks are in the same order, -1 if they are in different orders. We may say that we score $+1$ or -1 according as the pair agree or disagree in the two rankings.

The same procedure is followed for each possible pair from the ranking of 10. There are 45 such pairs and the scores are as follows (we write them down in full so that the reader can follow the method, but in practice, as we show presently, this is unnecessary):

Pair	Score	Pair	Score	Pair	Score	Pair	Score
AB	−1	BF	−1	DE	+1	FH	−1
AC	+1	BG	−1	DF	+1	FI	+1
AD	+1	BH	−1	DG	+1	FJ	−1
AE	+1	BI	−1	DH	+1	GH	+1
AF	−1	BJ	−1	DI	+1	GI	−1
AG	+1	CD	+1	DJ	+1	GJ	+1
AH	−1	CE	−1	EF	−1	HI	−1
AI	−1	CF	−1	EG	+1	HJ	−1
AJ	+1	CG	+1	EH	+1	IJ	−1
BC	+1	CH	−1	EI	−1		
BD	+1	CI	−1	EJ	−1		
BE	−1	CJ	+1	FG	−1		

The total of positive scores, say P, is 21 and that of negative scores, say $-Q$, is -24. Adding these two we arrive at a total score, say S, of -3.

Now if the rankings are identical each of the 45 unit scores will be positive and hence the maximum value of S is 45. Similarly the minimum value of S is -45. We therefore calculate t as

$$\frac{\text{Actual score}}{\text{Maximum possible score}} = -\frac{3}{45} = -0.07.$$

This value is close to zero and indicates very little correlation between the two rankings. A zero value may, in fact, be regarded as indicative of independence—halfway, so to speak, between complete positive dependence and complete negative dependence.

1.10 Consider now the general case. If we have two rankings of n, the number of pairs of comparisons which can be made is equal to the number of ways of choosing two things from n, which is $\frac{1}{2}n(n-1)$, sometimes written as $\binom{n}{2}$. This is the maximum value of the score, attained when the rankings are identical. If S is the total score we define the correlation coefficient by

$$t = \frac{2s}{n(n-1)}. \tag{1.3}$$

If P and Q are the totals of the positive and negative scores respectively we have the equivalent forms since $P + Q = \frac{1}{2}n(n-1)$

$$t = \frac{2(P-Q)}{n(n-1)} \tag{1.4}$$

$$= 1 - \frac{4Q}{n(n-1)} \tag{1.5}$$

$$= \frac{4P}{n(n-1)} - 1. \tag{1.6}$$

If there is complete agreement between the rankings, $Q = 0$ and $t = 1$, its maximum value; similarly, $P = 0$ and $t = -1$ if there is complete disagreement. If $P = Q$, the agreement and disagreement balance out and $t = 0$.

1.11 The determination of the score S (or equivalently of P or of Q) does not require the detailed procedure we have followed above. There are several short-cut methods of which the following are probably the easiest.

(a) Consider the rearranged form of (1.2). When one ranking is in the natural order, $1, 2, \ldots, n$, all unit scores arising from it are positive. Consequently, the contributions to P will arise only from pairs in the second ranking which are in the natural or direct order. These are all we need to count. The second ranking is

8   9   3   7   4   1   5   2   6   10

Considering first the pairs associated with the first member 8, we see that there are two members greater than 8 on the right of it. The contribution to P is therefore $+2$. Taking now pairs associated with 9 (other than 8 9 which has already been taken into account) we find a contribution to P of $+1$. Similarly the contribution of pairs associated with 3, arising from members to the right of it, is $+5$. Proceeding in this way we find

$$P = 2 + 1 + 5 + 1 + 3 + 4 + 2 + 2 + 1 = 21.$$

Hence, from (1.6),

$$t = \frac{42}{45} - 1 = -0.07, \text{ as before.}$$

(b)   If it is too troublesome to rearrange the rankings so that one of them is in the natural order we may proceed as follows. In the rankings from (1.1) write the natural order above them as:

	1	2	3	4	5	6	7	8	9	10
X	7	4	3	10	6	2	9	8	1	5
Y	5	7	3	10	1	9	6	2	8	4

The number 1 in ranking Y has a 6 above it in ranking X. In the natural ranking 6 has four members to the right. Score 4 and delete the 6 from the natural ranking. Now the number 2 in Y has an 8 above it in X and in the natural ranking 8 has two members to the right. Score 2 and strike out the 8 from the natural ranking. Proceeding in this way we find scores of

$$4 + 2 + 5 + 3 + 2 + 1 + 1 + 2 + 1 + 0 = 21,$$

which gives the value of P, as found before.

The validity of this rule is evident if we rearrange the rankings so as to put Y in the natural order (this being the order in which we have considered the members):

	5	8	3	10	1	7	2	9	6	4
X	6	8	3	5	7	9	4	1	2	10
Y	1	2	3	4	5	6	7	8	9	10

The contributions to P by method (b) are the same as those given by method (a) applied to the rankings Y and X. There are, for instance, 4 members to the right of 6 in X which are greater than 6; 2 members to the right of 8 in X which are greater than 8; and so on.

It may help to give some idea of the values assumed by t in particular cases if we set out some rankings of 10 and the corresponding t obtained by correlating them with the natural order. The reader should check these values as an exercise.

Ranking											Value of t
a	4	7	2	10	3	6	8	1	5	9	+0.11
b	1	6	2	7	3	8	4	9	5	10	+0.56
c	7	10	4	1	6	8	9	5	2	3	−0.24
d	6	5	4	7	3	8	2	9	10	1	+0.02
e	10	1	2	3	4	5	6	7	8	9	+0.60
f	10	9	8	7	6	1	2	3	4	5	−0.56

Tau as a coefficient of concordance

1.12   For any two pairs of ranks (x_i, y_i) and (x_j, y_j) out of the n, define them as *concordant* if $y_i < y_j$ when $x_i < x_j$ or $y_i > y_j$ when $x_i > x_j$, or equivalently if $(x_i - x_j)(y_i - y_j) > 0$. Similarly we define any two pairs as *discordant* if $y_i < y_j$ when $x_i > x_j$ or $y_i > y_j$ when $x_i < x_j$, or equivalently if $(x_i - x_j)(y_i - y_j) < 0$. Therefore, each pair with a score of $+1$ is called concordant and each pair with a score of -1 is called discordant. Thus P is the number of concordant pairs, Q is the number of discordant pairs, and S is the excess of concordant pairs over discordant pairs. Further, since the total number of pairs is $n(n - 1)/2$, t as defined by (1.4) is the proportion of concordant pairs minus the proportion of discordant pairs; t is therefore a relative measure of concordance between the two sets of n rankings.

Tau as a coefficient of disarray

1.13   The coefficient as we have introduced it provides a kind of average measure of the agreement between pairs of objects ('agreement', that is to say, with respect to order) and thus has evident recommendations as a measure of concordance between the two rankings. There is another instructive way of looking at the coefficient. Consider the two rankings of 7:

$$
\begin{array}{cccccccc}
X & 1 & 2 & 3 & 4 & 5 & 6 & 7 \\
Y & 6 & 3 & 5 & 7 & 1 & 2 & 4
\end{array}
$$

We may transform Y into X by successively interchanging pairs of neighbours in Y. For instance, interchanging 1 with its left-hand neighbours, we have, in four stages:

$$
\begin{array}{ccccccc}
6 & 3 & 5 & 1 & 7 & 2 & 4 \\
6 & 3 & 1 & 5 & 7 & 2 & 4 \\
6 & 1 & 3 & 5 & 7 & 2 & 4 \\
1 & 6 & 3 & 5 & 7 & 2 & 4
\end{array}
$$

Now interchanging 2 we find, after four more stages,

$$
\begin{array}{ccccccc}
1 & 2 & 6 & 3 & 5 & 7 & 4.
\end{array}
$$

Interchanging the 3 and 6 gives us

$$
\begin{array}{ccccccc}
1 & 2 & 3 & 6 & 5 & 7 & 4.
\end{array}
$$

Interchanging with the 4 gives us, in three stages,

$$
\begin{array}{ccccccc}
1 & 2 & 3 & 4 & 6 & 5 & 7.
\end{array}
$$

Finally, interchanging the 6 and 5 gives us the natural order X.

This transformation has taken 13 moves, and we could not have made it in fewer. We might have taken more, as for instance if we had interchanged 1 and 2 and back again before making the above sequence of moves. It will be clear that there is a minimum number of moves which transform any ranking into any other ranking of the same number of members. Denote this minimum number of interchanges by s.

Then we shall show in the next chapter that

$$s = Q \text{ or equivalently}$$

$$s = \tfrac{1}{2}\{\tfrac{1}{2}n(n - 1) - S\} \tag{1.7}$$

which gives a simple relation between the minimum number of interchanges s and the number of negative scores Q or the total score S. In the example we have just considered $S = -5$, $n = 7$, and hence

$$s = \tfrac{1}{2}(21 + 5) = 13, \text{ as found.}$$

From (1.5) and (1.7) it follows that

$$t = 1 - \frac{4s}{n(n-1)} \tag{1.8}$$

exhibiting t as a simple function of the minimum number of interchanges between neighbours required to transform one ranking into the other—in short, as a kind of coefficient of disarray.

Spearman's rho

1.14   We now discuss another coefficient of rank correlation denoted by r_s and called Spearman's rho after C. Spearman, who introduced it into psychological work. Consider again the two rankings of 10 given in (1.1):

Math:	7	4	3	10	6	2	9	8	1	5
Music:	5	7	3	10	1	9	6	2	8	4
Differences d	2	-3	0	0	5	-7	3	6	-7	1
Differences d^2	4	9	0	0	25	49	9	36	49	1

We have subtracted the ranks for music from those for math and shown the results in the row labelled 'Differences d'. These differences should sum to zero (which provides an arithmetical check) because the sum is the difference of two quantities each of which is the sum of the numbers 1 to 10. We have also shown the squares of these differences. Denoting now the sum of these squares by $\sum d^2$ we define Spearman's rho by the equation

$$r_s = 1 - \frac{6 \sum d^2}{n^3 - n} \tag{1.9}$$

or, in our present example,

$$r_s = 1 - \frac{6(182)}{990}$$

$$= -0.103.$$

1.15   When two rankings are identical all the differences d are zero and from (1.9) it follows that $r_s = 1$. We will now prove that when one ranking is the reverse of the other, $r_s = -1$.

Suppose that n is odd, and is equal to $2m + 1$. We lose no generality by writing one ranking in the natural order, and the rankings and differences may then be expressed as follows:

X:	1,		2, ..., m,	$m + 1$, $m + 2$, ..., $2m$,	$2m + 1$
Y: $2m + 1$,		$2m$, ..., $m + 2$,	$m + 1$,	m, ..., 2,	1
d:	$-2m$, $-(2m - 2)$, ..., -2,		0,	2, ..., $2m - 2$,	$2m$

$$\tag{1.10}$$

The sum of squares is thus given by

$$\sum d^2 = 8[m^2 + (m-1)^2 + \cdots + 2^2 + 1^2]$$

$$= 8m(m+1)(2m+1)/6$$

$$= (n-1)(n+1)(n)/3$$

$$= (n^3 - n)/3.$$

If we substitute this value in (1.9) we find

$$r_s = 1 - 6/3 = -1.$$

If n is even, say equal to $2m$, we have similarly

X:	1,		2, ..., m,	m + 1, ..., 2m − 1,		m
Y:		2m,	2m − 1, ..., m + 1,	m, ...,	2,	1
d: −(2m − 1),	−(2m − 3), ..., −1,			1, ..., 2m − 3, 2m − 1		

$$(1.11)$$

Thus

$$\sum d^2 = 2[(2m-1)^2 + (2m-3)^2 + \cdots + 3^2 + 1^2]$$

$$= 2[(2m)^2 + (2m-1)^2 + (2m-2)^2 + \cdots + 3^2 + 2^2 + 1^2]$$

$$\quad - 2[(2m)^2 + (2m-2)^2 + \cdots + 2^2]$$

$$= 4m(2m+1)(4m+1)/6 - 8m(m+1)(2m+1)/6$$

$$= (n^3 - n)/3$$

and, as before, if we substitute in (1.9) we find $r_s = -1$.

The coefficient r_s can thus attain the values -1 and $+1$. We shall show in Chapter 2 that r_s cannot lie outside that range and assumes the extreme values only when there is perfect disagreement or agreement between the rankings.

1.16 The reason for taking the sum of *squares* of the rank-differences will be clear to the reader who is familiar with the calculation of statistical measures of dispersion such as the standard deviation. It is obvious enough that we cannot base a coefficient on the sum of differences $\sum d$, for this is zero. It might be thought, however, that if we disregarded the signs of the differences and summed them a somewhat simpler coefficient could be reached; and indeed this was one of Spearman's original suggestions. The procedure leads to difficulties in the more advanced theory, particularly in connection with sampling questions, and we shall not pursue it.

1.17 The score Q of equation (1.5) is simply the number of pairs which occur in different orders in the two rankings. We may call any such case an 'inversion' and t is thus a linear function of the number of inversions. It is interesting to observe that r_s can also be regarded as a coefficient of inversion when each inversion is weighted; in fact, if the pair of ranks (i, j) are inverted $(i < j)$ and we score $(j - i)$ for any inversion, and if we sum all such scores to obtain a total V, we have

$$V = \sum d^2/2 \qquad (1.12)$$

and hence

$$r_s = 1 - \frac{12V}{(n^3 - n)}. \qquad (1.13)$$

We shall prove this result in Chapter 2.

For example, consider the two rankings of 7 from Section 1.13, namely

$$
\begin{array}{cccccccc}
X & 1 & 2 & 3 & 4 & 5 & 6 & 7 \\
Y & 6 & 3 & 5 & 7 & 1 & 2 & 4
\end{array}
\qquad (1.14)
$$

Taking those pairs of ranks which are inverted in the second ranking as compared with the first, we have

Ranks inverted	Weight
6 3	3
6 5	1
6 1	5
6 2	4
6 4	2
3 1	2
3 2	1
5 1	4
5 2	3
5 4	1
7 1	6
7 2	5
7 4	3
	40

The number of inversions is 13, as we found in Section 1.13, and

$$t = -5/21 = -0.24.$$

The sum of weighted inversions is $V = 40$ and hence

$$r_s = 1 - \frac{480}{336}$$

$$= -0.43.$$

It is easily verified from (1.13) that for the rankings (1.14) we have $\sum d^2 = 80$, which equals $2V$ as (1.12) states in general.

Conjugate rankings

1.18 One property which is common to both t and r_s may be noticed. If we correlate a given ranking X with the natural order $Y(1, \ldots, n)$, and again with the inverse order Y' $(n, \ldots, 1)$, the values of t are the same in magnitude but opposite in sign. This will be seen from the definition of t, for the effect of reversing the order Y is to reverse the sign of each unit score contributing to S, and hence the sign of S itself. Thus, corresponding to any two rankings X and Y (not necessarily natural orders) with correlation t there will be rankings X and Y' with correlation $-t$. Consider, for instance, the two rankings of 7:

$$
\begin{array}{cccccccc}
X & 4 & 1 & 6 & 7 & 5 & 2 & 3 \\
Y & 7 & 6 & 3 & 1 & 4 & 5 & 2
\end{array}
\qquad (1.15)
$$

Let us rearrange X in the natural order. We then have

$$
\begin{array}{cccccccc}
X & 1 & 2 & 3 & 4 & 5 & 6 & 7 \\
Y & 6 & 5 & 2 & 7 & 4 & 3 & 1
\end{array}
$$

and the value of t is readily found to be $-11/21$ or -0.52. If we now invert the natural order X to obtain

$$
\begin{array}{cccccccc}
X' & 7 & 6 & 5 & 4 & 3 & 2 & 1 \\
Y & 6 & 5 & 2 & 7 & 4 & 3 & 1
\end{array}
$$

we find the correlation of X' and Y is $+0.52$. If we then rearrange so that Y is in the same order as (1.15), we obtain

$$
\begin{array}{cccccccc}
X' & 4 & 7 & 2 & 1 & 3 & 6 & 5 \\
Y & 7 & 6 & 3 & 1 & 4 & 5 & 2
\end{array}
\qquad (1.16)
$$

The rankings X and X' may be regarded as conjugate. When correlated with Y they give values of t which are equal in magnitude but opposite in sign and this is true in general.

1.19 It is also true, though not so obvious, that the values of r_s between X, Y and X', Y are equal in magnitude but opposite in sign. We shall prove the general result in the next chapter. The reader may verify as an exercise that for the particular cases (1.15) and (1.16) we have respectively $r_s = -17/28$ and $r_s = +17/28$.

1.20 In a sense, therefore, the scales of measurement of rank correlation set up by the use of t or r_s are symmetrical about the value zero. They range from -1 to $+1$, and corresponding to any given positive value of t or of r_s there is a negative value of the same magnitude arising from an inversion of one of the rankings. The scales, we may say, are unbiased.

1.21 The reader must not expect to find that the numerical values of t and r_s are the same for any given pair of rankings, except when there is complete agreement or disagreement. The rankings given at the end of Section 1.12

when correlated with the natural order produce the following values:

Ranking	t	r_s
a	+0.11	+0.14
b	+0.56	+0.64
c	−0.24	−0.37
d	+0.02	+0.03
e	+0.60	+0.45
f	−0.56	−0.76

These will illustrate the sort of diffferences in magnitude which arise in practice. The coefficients have different scales, like those of the Celsius and Fahrenheit thermometers; and they differ in rather more than scale, since r_s gives greater weight to inversions of ranks which are farther apart. In practice we often find that, when neither coefficient is too close to unity, r_s is about 50 per cent greater than t in absolute value, but this is not an invariable rule.

Daniels' inequality

1.22   It is possible to state certain inequalities connecting the values which r_s and t can take for a given ranking of n objects. The first one, due to Daniels, is as follows:

$$-1 \le \frac{3(n + 2)}{n - 2} t - \frac{2(n + 1)}{n - 2} r_s \le 1. \tag{1.17}$$

For large n this becomes effectively

$$-1 \le 3t - 2r_s \le 1. \tag{1.18}$$

If t is greater than zero the upper limit may be attained but not the lower limit; when t is less than zero the lower limit may be attained but not the upper limit; and when $t = 0$ both limits may be attained. Thus, we could have a ranking for which $t = 0$ and $r_s = 0.5$. It would, however, be a rather peculiar one. Consider, for instance, the rankings

$$\begin{array}{ccccccccc}
X & 1 & 2 & 3 & 4 & 5 & 6 & 7 & 8 \\
Y & 5 & 6 & 7 & 8 & 1 & 2 & 3 & 4
\end{array} \tag{1.19}$$

We find

$$r_s = -0.53, \qquad t = -0.14.$$

The Durbin–Stuart inequality

1.23   Durbin and Stuart have found other inequalities satisfied by r_s and t which enable us to set upper and lower limits to r_s for given t. They showed that, for any ranking, the sum of weighted inversions V of Section 1.17 is related to Q by

$$V \ge \tfrac{2}{3}Q\left(1 + \frac{Q}{n}\right), \tag{1.20}$$

the equality being attainable in certain cases. By using (1.13) and (1.5) we then find

$$r_s \le 1 - \frac{1 - t}{2(n + 1)}\{(n - 1)(1 - t) + 4\}, \qquad t \ge 0, \qquad (1.21)$$

which gives an upper limit for r_s as a function of any given positive t. A lower limit is given by (1.17) as

$$r_s \ge \frac{3nt - (n - 2)}{2(n + 1)}, \qquad t \ge 0. \qquad (1.22)$$

For n large these limits give us

$$\tfrac{3}{2}t - \tfrac{1}{2} \le r_s \le \tfrac{1}{2} + t - \tfrac{1}{2}t^2, \qquad t \ge 0 \qquad (1.23)$$

both limits being attainable. For example when $t = 0$, $-\tfrac{1}{2} \le r_s \le \tfrac{1}{2}$. When $t = \tfrac{1}{2}$, $\tfrac{1}{4} \le r_s \le \tfrac{7}{8}$; when $t = 0.9$, $0.85 \le r_s \le 0.995$.

For $t < 0$ we find

$$\tfrac{1}{2}t^2 + t - \tfrac{1}{2} \le r_s \le \tfrac{3}{2}t + \tfrac{1}{2}. \qquad (1.24)$$

We shall prove these results in Chapter 2. They illustrate the general remark, made above, that although r_s and t are related the relationship is not a simple one.

Stragglers

1.24   It sometimes happens that when a ranking has been made, new individuals are added which necessitate re-ranking. Similarly, in writing down the ranks of a series of individuals which are distinguished by variate-values or marks but are in disorder, we may make mistakes and find at the end of the ranking that a few have been omitted. The addition of members to a ranking does not require a complete re-calculation of either t or r_s. An example will make the point clear.

Example 1.1

A confidential inquiry is sent out to a number of firms asking for the rate of dividend which they propose to declare at their next annual general meetings. We will suppose that they are all able to answer this question, but that there is some possibility that those with the higher dividends in prospect are more reluctant to reply and will delay or will not reply at all. We will also assume that all the rates are different. These are perhaps not very realistic assumptions but they will serve the purpose of this example.

By a certain date a number of replies have been received and it is then necessary to close the inquiry and to summarise the results. How far can we assume that the replies to date are representative of the population to which the inquiry was addressed? Is there any evidence to suggest that those who reply earlier differ from those who reply later?

Suppose we receive 15 replies in the following order:

(X) Order of receipt:	1	2	3	4	5	6	7	8	9	10	11	12	13	14	15
(Y) Percentage dividend:	15	13	12	16	25	8	9	14	17	11	18	20	10	21	19
(Z) Rank of percentage:	8	6	5	9	15	1	2	7	10	4	11	13	3	14	12

If there *is* some relation between the order of receipt and the magnitude of the percentage, it ought to be evidenced by the correlation between the order of receipt and the order of the percentages according to magnitude shown in the last row above. For the correlation of X and Z we find $S = 25$,

$$t = \frac{25}{105} = 0.24.$$

We also find $\sum d^2 = 392$,

$$r_s = 1 - \frac{6(392)}{3360} = +0.30.$$

This suggests some small positive correlation between X and Z, and in Chapter 4 we shall see how to test its significance. That, however, is not the point of the present example. Suppose that, after these values of t and r_s have been worked out, two more replies arrive with percentages 7 and 23. Nearly all the ranks in Z are affected and have to be re-numbered.

However, the effect of the addition of the two extra values on S can be ascertained very simply. The new member with percentage 7 and rank 16 in the X-ranking merely has to be considered in relation to the other fifteen members, and since it has a lower percentage than any of them it adds -15 to the score S. Similarly, the new member with percentage 23 adds 14. The new score S is therefore $14 - 15 = -1$ more than the old, i.e. is 24, and the new value of t is given by

$$t = \frac{24}{136} = +0.18.$$

In this way a kind of running total of t can be ascertained without the necessity of re-ranking at each stage.

The effect of the two additional values on $\sum d^2$ and therefore r_s can be determined with only a little more effort using the relation in (1.12) that shows $\sum d^2$ as two times the sum of the weighted inversion scores. We must add to our previous $\sum d^2$ twice the sum of the differences of ranks for all pairs that the new observation is discordant with.

The first additional observation is a percentage 7 which corresponds to an X rank of 16. Since 7 is smaller than every other percentage Y, it is discordant with the earlier responses which had X ranks $1, 2, \ldots, 15$. The addition to the sum of weights for these 15 inversions is then

$$(16 - 1) + (16 - 2) + (16 - 3) + \cdots + (16 - 15) = 120,$$

which adds $2(120) = 240$ to the old $\sum d^2$. The second additional observation is a percentage 23 which will be paired with an X rank of 17. Because percentage 23 is smaller than only percentage 25 which had an X rank of 5, the addition to the sum of weighted inversion scores is $(17 - 5) = 12$, which adds $2(12) = 24$ to $\sum d^2$. The new $\sum d^2$ based on all 17 observations is then $392 + 240 + 24 = 656$ and the new value of r_s is

$$r_s = 1 - \frac{6(656)}{4896} = 0.196.$$

It may be remarked that in this example the ranks of ranking Z are obtained from a variate, the percentage stated in the reply. Those of the order of receipt are not obtained from a variate, although we could, with sufficient patience, measure the time-intervals elapsing between the receipt of consecutive replies and hence regard the X ranking as arranged according to a time-scale.

This is therefore an example of what is frequently called time-series data. Both tau and rho are useful in describing relationships for data which consist of a single set of variate values whose order is according to time (or space) intervals. If the time order is from earliest to latest and the variate rankings are from smallest to largest a large positive relation indicates a positive trend over time and a strong negative relation indicates a negative trend. Significance tests for trend based on tau and rho will be covered in Chapter 4.

1.25 We conclude with three further examples of the use of tau and rho to measure rank correlations.

Example 1.2

Twelve similar discs are constructed, ranging from light blue to dark blue in colour. Their order is known objectively by a colorimetric test. To test the ability of a dress designer to distinguish shades she is shown these discs and asked to arrange them in order. The results are as follows:

Objective order:	1	2	3	4	5	6	7	8	9	10	11	12
Order assigned by the subject:	1	4	7	2	3	5	8	12	10	6	11	9

We want to measure the subject's ability to distinguish the different shades of blue.

For the value of P we have

$$11 + 8 + 5 + 8 + 7 + 6 + 4 + 0 + 1 + 2 + 0 = 52$$

$$t = \frac{104}{66} - 1 = +0.58.$$

The correlation is positive and substantial, but far from perfect. We shall show how to test its significance in Chapter 4.

In this example we are measuring the agreement between a subjective and an objective order. The subject's failure to achieve complete success may be due to genuine inability to distinguish finer shades, to wandering attention or to other causes; but whatever the cause we can test the subject's ability against a given objective order.

Example 1.3

Consider now the case where three judges rank a number of competitors in a talent contest as follows:

Judge X:	1	2	3	4	5	6	7	8	9
Judge Y:	5	4	1	7	2	8	3	6	9
Judge Z:	2	5	1	3	4	7	6	9	8

There is here no objective order such as existed in the previous example. We are interested in how closely the judges agree among themselves, not in their agreement with some objective standard.

Therefore we will compute the tau coefficient between each pair of judges. For this purpose, the three pairs of rankings are shown in Table 1.1 in a convenient form for calculation of P and Q. Note that the rankings are rearranged so that the ranks of the left member of each pair of judges occur in natural order.

These results indicate that judges X and Z agree to a greater extent than judges X and Y or judges Y and Z. Judges X and Y show the least agreement.

Table 1.1   Computation of Tau for All Pairs of Judges

Judges X and Y				Judges X and Z				Judges Y and Z			
X	Y	P	Q	X	Z	P	Q	Y	Z	P	Q
1	5	4	4	1	2	7	1	1	1	8	0
2	4	4	3	2	5	4	3	2	4	5	2
3	1	6	0	3	1	6	0	3	6	3	3
4	7	2	3	4	3	5	0	4	5	3	2
5	2	4	0	5	4	4	0	5	2	4	0
6	8	1	2	6	7	2	1	6	9	0	3
7	3	2	0	7	6	2	0	7	3	2	0
8	6	1	0	8	9	0	1	8	7	1	0
9	9	—	—	9	8	—	—	9	8	—	—
		24	12			30	6			26	10

The tau coefficients are

$$t(X \text{ and } Y) = \frac{2(24 - 12)}{9(8)} = 0.33$$

$$t(X \text{ and } Z) = \frac{2(30 - 6)}{9(8)} = 0.67$$

$$t(Y \text{ and } Z) = \frac{2(26 - 10)}{9(8)} = 0.44$$

Example 1.4

Table 1.2 shows the number of ocean-going steam and motor ships and the total gross tons of those ships for 19 countries as of 1 July 1986. The table also shows the ranks of these countries according to these two variates. The countries are rearranged from smallest to largest according to number of ships in Table 1.3 to facilitate the calculation of tau and rho. The results are

$$r_s = 1 - \frac{6(258)}{6137} = 0.75$$

$$t = \frac{2(135 - 36)}{19(18)} = 0.58.$$

These values seem to express fairly well the relationships between the variables. The countries with larger numbers of ships are, on the whole, those with the greater total weight of ships, but Norway, British Columbia and Germany form notable exceptions and reduce the strength of relationship for all countries simultaneously.

The reader may verify that the Pearson product-moment correlation coefficient of ordinary statistical theory for the original variate values in Table 1.2 is equal to 0.75, which is the same as the Spearman rank correlation coefficient. It appears that in this example we have not altered the pertinent information by replacing the variate values by ranks.

In some cases, especially with economic data where the magnitudes differ widely, the use of variate values may present a distorted view of the relationship by letting one or two extremely large numbers essentially eliminate the

Table 1.2 Number and Total Weight of Oceangoing Steam and Motor Ships as of 1 July 1986

Country	Number	Total gross tons
British Columbia	450	9 825
China (People's Republic)	1 025	10 278
Cyprus	716	8 900
Germany (West)	528	4 931
Greece	1 835	32 092
India	358	6 434
Italy	569	7 855
Japan	1 604	37 366
Korea	487	6 575
Liberia	1 852	62 126
Netherlands	464	4 361
Norway	424	12 572
Panama	3 620	42 101
Philippines	399	4 892
Singapore	480	6 652
Spain	489	5 191
U.S.S.R.	2 514	18 717
United Kingdom	541	12 744
United States	737	16 024

Table 1.3 Calculations Based on Table 1.2

Country	Rank by number	Rank by tons	d^2	P	Q
India	1	5	16	14	4
Philippines	2	2	0	16	1
Norway	3	12	81	7	9
British Columbia	4	10	36	8	7
Netherlands	5	1	16	14	0
Singapore	6	7	1	10	3
Korea	7	6	1	10	2
Spain	8	4	16	10	1
Germany	9	3	36	10	0
United Kingdom	10	13	9	6	3
Italy	11	8	9	8	0
Cyprus	12	9	9	7	0
United States	13	14	1	5	1
China	14	11	9	5	0
Japan	15	17	4	2	2
Greece	16	16	0	2	1
Liberia	17	19	4	0	2
U.S.S.R.	18	15	9	1	0
Panama	19	18	1		
			258	135	36

effects of small ones. Then replacing the variate values by ranks tends to restore a balance and give each country an equal voice, as it were, in the discussion. Whether or not this is a sound procedure depends on circumstances. The point is that sometimes the variate values can be more misleading than the ranks.

1.26 Another use of rank correlation arises when data are collected as opinions expressed on a Likert scale for two different groups. A five-point Likert scale for example may request subjects to read a statement and respond as some integer between 1 and 5 where 1 = strongly disagree, 2 = moderately disagree, 3 = neutral or no opinion, 4 = moderately agree, 5 = strongly agree. The scale may be reduced to only 3 points such as disagree, no opinion and agree, or it may be elongated to more than 5 points. Of course, the response need not be an opinion. It could relate to perceived importance, degree of support, attitude or any other subjective evaluation. Likert scale surveys are conducted frequently in the social and behavioral sciences. If we have Likert scale data on a single question as responses by two different groups of m subjects, the limited range of responses makes it very difficult, if not impossible, to rank from 1 to m for each group. Besides, in most surveys there is not just one question but several questions relating to a single topic. In such cases the usual procedure is to find the average of the Likert scale responses to each question by each group. The resulting average responses for all questions can be ranked from 1 to n if there are n questions and then tau or rho can be calculated in the usual way for the two groups. Note that here n is the number of questions, not the number of subjects or respondents. The following example illustrates these procedures with averages of Likert scale data.

Example 1.5

Davis (1970) explored the extent to which husbands and wives agree in their perception of the role each partner plays in their domestic decision making. In a sample of 97 couples, each individual was asked to respond to six questions concerning the automobile most recently purchased. The questions addressed the issues of (a) when to purchase, (b) how much to spend, (c) make, (d) model, (e) colour, and (f) where to purchase. Each of these questions was rated on a 5-point scale with (1) husband decided, (2) husband had more influence, (3) equal influence, (4) wife had more influence than husband, (5) wife decided. The average scores of the 97 couples are shown in Table 1.4. Describe the relationship of perceptions between husbands and wives using both tau and rho.

Table 1.4

Question	Husbands	Wives
(a)	1.95	1.83
(b)	1.97	1.95
(c)	2.05	1.98
(d)	2.13	2.11
(e)	2.41	2.17
(f)	2.95	2.73

The first step is to rank the average ratings from 1 to 6 as shown in Table 1.5.

Table 1.5

Question	Husbands' rank	Wives' rank	d^2	P	Q
(a)	1	1	0	5	0
(b)	2	2	0	4	0
(c)	3	3	0	3	0
(d)	4	4	0	2	0
(e)	5	5	0	1	0
(f)	6	6	0	9	9
			0	15	0

$$t = \frac{2(15 - 0)}{6(5)} = 1.00$$

$$r_s = 1 - \frac{6(0)}{6(35)} = 1.00$$

The agreement of perceptions is perfect with each descriptive measure.

Notes and references

The coefficient tau was considered by Greiner (1909) and Esscher (1924) as a method of estimating correlations in a normal population (see Chapters 9 and 10). Lindeberg (1925, 1929) also discussed this measure and gave its variance. Tau was rediscovered purely as a coefficient of rank correlation by Kendall (1938). Rosander (1942) independently gave a test based on inversions that is equivalent to tau. Kruskal (1958) traced the history of the introduction of tau. For tau as a coefficient of disarray, see Haden (1947), Feller (1945) and Moran (1947).

The early references for Spearman's rho are Spearman (1904, 1906), Pearson (1907) and Kendall, Kendall and Smith (1939). Hotelling and Pabst (1936) trace the history of rho.

Early papers dealing with the computation of tau are Bright (1954) and Lieberson (1961). A computer method of calculation was developed by Knight (1966). A graphic computation of tau as a coefficient of disarray was presented by Griffin (1958) and modified by Shah (1961).

Noether (1981) argued that the interpretation of Kendall's tau as a coefficient of concordance is simple and intuitive and therefore has practical and pedagogical advantages over other measures of rank correlation. Wilkie (1980) gave a pictorial representation of the number of discordant pairs. The interpretation of Spearman's rho is much more difficult as is evidenced by Kruskal (1958) and Griffiths (1980). A mechanical interpretation of rho was given by Evans (1973). Nevertheless, Spearman's coefficient is probably better known than Kendall's and possibly more frequently used. Geometric interpretations of both rho and tau were given in Schulman (1979).

Papaioannou and Speevak (1977) gave upper bounds for r_s that are applicable when new observations are added, a situation considered in Section 1.24. Similar inequalities were given for tau. A reversal of the inequalities gives lower bounds applicable when observations are missing or deleted. Additional inequalities for missing and added observations were obtained by Papaioannou and Loukas (1984) for rho, tau, Spearman's footrule (see Chapter 2) and the coefficient of concordance (see Chapter 6).

The expressions for $\sum d^2$ and r_s in (1.12) and (1.13) as a function of the inversion weights were given in Durbin and Stuart (1951). Schulman (1979) used this result to show how to update r_s for additional observations; this method was illustrated in Example 1.1.

The inequalities and bounds given in (1.17)–(1.24) that relate r_s and t were developed in Daniels (1950; 1951) and Durbin and Stuart (1951) and are proved here in Chapter 2. Some of these basic results were extended by Papaioannou and Speevak (1977) to the case of missing or deleted observations and also to additional observations.

Problems, Fischhoff and Lichtenstein

1.1 Slovic (1980) reported a study designed to see whether experts and non-experts have the same judgment concerning the relative frequency and risk of death from 30 different hazardous activities, substances and technologies. For one aspect of the study, the non-expert subjects used were members of the League of Women Voters in Eugene, Oregon; this group was selected to represent the opinions of educated, informed citizens. The experts used were persons across the United States who are professionally involved in assessing risks. The non-experts agreed with the experts about risks on items like power mowers, motor vehicles and handguns, but the groups disagreed about nuclear power and X-rays. The overall rankings are shown in Table 1.6 where rank 1 represents a judgment of most risk and 30 of least risk.

(a) Calculate tau for these data
(b) Calculate rho for these data.

Table 1.6

Activity	Non-experts	Experts
Nuclear power	1	20
Motor vehicles	2	1
Handguns	3	4
Smoking	4	2
Motorcycles	5	6
Alcoholic beverages	6	3
General (private) aviation	7	12
Police work	8	17
Pesticides	9	8
Surgery	10	5
Firefighting	11	18
Large construction	12	13
Hunting	13	23
Spray cans	14	26
Mountain climbing	15	29
Bicycles	16	15
Commercial aviation	17	16
Electric power (non-nuclear)	18	9
Swimming	19	10
Contraceptives	20	11
Skiing	21	30
X-rays	22	7
High school and college football	23	27
Railways	24	19
Food preservatives	25	14
Food colouring	26	21
Power mowers	27	28
Prescription antibiotics	28	24
Home appliances	29	22
Vaccinations	30	25

Answers: $r_s = 0.5933$, $t = 0.4391$

1.2 A health agency is investigating the relationship between air pollution and prevalence of pulmonary related disease. The data in Table 1.7 are

ranks (1 = worst, 10 = best) of ten cities on these two characteristics.
Describe the relationship using both rho and tau.

Table 1.7

City	Pollution	Disease
A	4	5
B	7	4
C	9	7
D	1	3
E	2	1
F	10	10
G	3	2
H	5	8
I	6	6
J	8	9

Answers: $r_s = 0.8182$, $t = 0.6444$

1.3 The research in Davis (1970) discussed in Example 1.5 also asked the 97
couples questions about purchasing furniture. The six questions were the
same as for cars except that (c) make became what kind of furniture,
(d) model became what style of furniture, and (e) colour became fabric
and colour. For the mean ranks in Table 1.8, compute tau and rho.

Table 1.8

Questions	Husbands	Wives
(a)	3.27	3.18
(b)	3.45	3.35
(c)	3.17	3.04
(d)	3.80	3.55
(e)	3.91	3.68
(f)	4.17	3.92

1.4 Hopwood and McKeown (1987) investigated whether the fact that
forecasts of earnings per share made by financial analysts are generally
more accurate than similar forecasts made by statistical models is due to
a timing advantage. The difference in timing is that statistical forecasts
rely only on last-quarter earnings and are generally made on the date of
the anouncement of past earnings while forecasts by analysts are often
made weeks later and can therefore take into account relevant recent
economic events. Relative accuracy is measured by the difference in
forecasting error, defined as absolute relative error by statistical model
forecast minus absolute relative error by financial analyst forecast, for 11
observations in one quarter taken from *Value Line*. Compute tau and
rho to describe the relationship for the data shown in Table 1.9.

Table 1.9

Observation	Difference in forecasting error	Number of days between forecasts
1	0.198	57
2	0.132	44
3	0.176	54
4	0.095	39
5	0.142	47
6	0.169	40
7	0.103	38
8	0.117	48
9	0.125	45
10	0.187	61
11	0.182	63

1.5 Dubinsky and Rudelius (1980–81) measured the relative importance of 11 different selling techniques for approaching a prospect by product salespeople and service salespeople. Samples of product and service personnel completed questionnaires where each selling technique was rated on a 5-point scale where 1 = no importance and 5 = extremely important. The data reported and shown in Table 1.10 are percentages of respondents giving a rating of very important or extremely important. Calculate tau and rho for the association between product and service salespeople.

Table 1.10

Selling technique for approaching a prospect	Product salespeople	Service salespeople
Survey approach	80%	59%
Question approach	75	71
Consumer benefit approach	73	78
Curiosity	56	77
Introductory approach	53	54
Referral approach	43	67
The compliment approach	34	50
Showmanship approach	21	23
Product approach	12	29
Shock approach	6	10
Premium approach	2	11

1.6 Wind, Mahajan and Swire (1983) compare various standardised portfolio models in several different ways. One part of the study gives ranks to 15 business units of a Fortune 500 multinational industrial firm that reflect their positions relative to two composite dimensions used as a basis for portfolio analysis. The first dimension is prospects for sector profitability (e.g. industry growth) and the second dimension is company's competitive capabilities (e.g. market share). Further, each ranking is computed by two different weighting schemes, first where the factors within each dimension are weighted equally and second where the factors are weighted unequally according to empirically derived measures of importance. The reported data are shown in Table 1.11. Calculate the tau and rho coefficients to measure the relation between the two weighting schemes for

(a)   the profitability dimension
(b)   the competitiveness dimension.

Table 1.11

Business unit	Profitability		Competitiveness	
	Rank with equal weights	Rank with unequal weights	Rank with equal weights	Rank with unequal weights
1	11	4	11	11
2	8	11	12	12
3	10	5	7	10
4	7	13	1	2
5	12	10	8	8
6	9	9	5	7
7	15	14	13	14
8	14	12	3	6
9	5	7	9	5
10	4	8	2	3
11	2	6	10	4
12	1	1	4	1
13	6	3	14	13
14	13	2	6	9
15	3	15	15	15

1.7   *American Health* (November 1987, Vol. VI, No. 9, p. 56) claimed that attributes commonly sought in a mate by both sexes are (A) adaptability, (B) creativity, (C) college graduate, (D) desire for children, (E) exciting personality, (F) good earning capacity, (G) good health, (H) good heredity, (I) good housekeeper, (J) intelligence, (K) kindness and understanding, (L) physical attractiveness and (M) religious orientation. Rank these 13 attributes in order of their importance to you, and ask your mate or friend of the opposite sex to do the same. Then calculate rho and tau as a measure of the agreement between you and your mate or friend.

1.8   David Buss, a psychologist at the University of Michigan, claimed that men's preferences of the attributes in Problem 1.7, in decreasing order of importance, are K, J, L, E, G, A, B, D, C, H, F, I, M, and that women's preferences are K, J, E, G, A, L, B, F, C, D, H, I, M. Measure the agreement between preferences of males and females and also between you and other members of your sex.

Chapter 2

Introduction to the general theory of rank correlation

2.1 In this chapter we shall begin the development of a general theory of rank correlation and shall demonstrate some results which were stated with ut proof in the previous chapter. Readers who are interested in practical applications and are prepared to take those results on trust can omit this chapter altogether; however, those readers with some previous knowledge of the theory of variate-correlation may profit from reading it to see how the various coefficients in current use may be linked together within the scope of a single theory.

The general correlation coefficient

2.2 Suppose we have a set of n objects which are being considered in relation to two properties represented by X and Y. Numbering the objects from 1 to n for the purposes of identification in any order we please, we may say that they exhibit values $x_1, \ldots, x_n$ according to X and $y_1, \ldots, y_n$ according to Y. These values may be variates or ranks.

To any pair of individuals, say the ith and the jth, we will allot an X-score, denoted by a_{ij}, subject only to the condition that $a_{ij} = -a_{ji}$. Similarly we will allot a Y-score, denoted by b_{ij}, where $b_{ij} = -b_{ji}$. We define a generalised correlation coefficient Γ by the equation

$$\Gamma = \frac{\sum a_{ij} b_{ij}}{\sqrt{(\sum a_{ij}^2 \sum b_{ij}^2)}}. \tag{2.1}$$

We regard a_{ij} as zero if $i = j$.

Tau as a particular case

2.3 This general definition includes t, r_s and the Pearson product-moment correlation r as particular cases which arise when particular methods of scoring are adopted.

Let p_i denote the rank of the ith object and p_j the rank of the jth object, both ranked according to the X quality. Suppose we allot a score $+1$ if $p_j > p_i$ (the ith member according to the X-quality) and -1 if $p_j < p_i$. Then

$$\begin{aligned} a_{ij} &= +1 \qquad p_i < p_j \\ &= -1 \qquad p_i > p_j \end{aligned} \tag{2.2}$$

25

and similarly for b and the Y scores. Thus the sum $\sum a_{ij}b_{ij}$ is equal to twice the sum S (twice because any given pair occurs once as (i, j) and once as (j, i) in the summation). Furthermore $\sum a_{ij}^2$ is merely the number of terms a_{ij}, that is, $n(n-1)$, and also for $\sum b_{ij}^2$. It follows from substitution in (2.1) that Γ is equal to the coefficient t as we defined it in Chapter 1.

r_s as a particular case

2.4   Instead of the simple ± 1 let us write

$$a_{ij} = p_j - p_i \tag{2.3}$$

and similarly

$$b_{ij} = q_j - q_i, \tag{2.4}$$

where q_i is the rank of the ith member according to the Y-quality. Both p_i and q_i range from 1 to n, and hence the sum of squares $\sum (p_j - p_i)^2$ and $\sum (q_j - q_i)^2$ are equal. From (2.1) we then have

$$\Gamma = \frac{\sum (p_j - p_i)(q_j - q_i)}{\sum (p_j - p_i)^2}. \tag{2.5}$$

Now

$$\sum_{i,j=1}^{n} (p_j - p_i)(q_j - q_i) = \sum_{i=1}^{n}\sum_{j=1}^{n} p_i q_i + \sum_{i=1}^{n}\sum_{j=1}^{n} p_j q_j$$

$$- \sum_{i=1}^{n}\sum_{j=1}^{n} (p_i q_j + p_j q_i)$$

$$= 2n \sum_{i=1}^{n} p_i q_i - 2 \sum_{i=1}^{n} p_i \sum_{j=1}^{n} q_j$$

$$= 2n \sum_{i=1}^{n} p_i q_i - \tfrac{1}{2}n^2(n + 1)^2, \tag{2.6}$$

since $\sum p_i$ and $\sum q_j$ are both equal to the sum of the first n positive integers, namely $\tfrac{1}{2}n(n + 1)$.

We also have

$$\sum d^2 = \sum_{i=1}^{n} (p_i - q_i)^2$$

$$= 2 \sum p_i^2 - 2 \sum p_i q_i, \tag{2.7}$$

and hence, from (2.6),

$$\sum (p_j - p_i)(q_j - q_i) = 2n \sum p_i^2 - \tfrac{1}{2}n^2(n + 1)^2 - n \sum d^2 \tag{2.8}$$

But $\sum p_i^2$ is the sum of squares of the first n positive integers, namely $\tfrac{1}{6}n(n + 1)(2n + 1)$, and the right-hand side of (2.8) thus reduces to

$$\tfrac{1}{6}n^2(n^2 - 1) - n \sum d^2. \tag{2.9}$$

Further

$$\sum (p_j - p_i)^2 = 2n \sum p_i^2 - 2 \sum p_i p_j$$
$$= 2n \sum p_i^2 - 2(\sum p_i)^2$$
$$= \tfrac{1}{6} n^2 (n^2 - 1) \tag{2.10}$$

and thus, on substituting from (2.9) and (2.10) in (2.5) we get

$$\Gamma = 1 - \frac{6 \sum d^2}{n^3 - n} \tag{2.11}$$

so that in this case Γ is reduced to Spearman's r_s.

Product-moment correlation as a particular case

2.5 Thirdly, suppose we base our scores on the actual variate-values and write

$$a_{ij} = x_j - x_i$$
$$b_{ij} = y_j - y_i. \tag{2.12}$$

Then

$$\tfrac{1}{2} \sum_{i,j} (x_j - x_i)(y_j - y_i) = n \sum_i x_i y_i - \sum_{i,j} x_i y_j, \tag{2.13}$$

$$\tfrac{1}{2} \sum_{i,j} (x_j - x_i)^2 = n \sum x_i^2 - (\sum x_i)^2. \tag{2.14}$$

Now the expression on the right-hand side of (2.13) is n times the covariance of X and Y, and that on the right-hand side of (2.14) is n times the variance of X. From (2.1) we then have

$$\Gamma = \frac{\text{cov}(X, Y)}{\sqrt{(\text{var } X \text{ var } Y)}}, \tag{2.15}$$

so that Γ becomes in this case the ordinary Pearson product-moment correlation of X and Y.

~ Spearman's r_s

2.6 It follows from the preceding paragraph that r_s itself may be regarded as a product-moment correlation between ranks considered as variates. We will verify this directly.

For a set of values which are the first n integers we have, as in the previous section,

$$\sum p_i = \tfrac{1}{2} n(n + 1)$$

and hence the first moment (the mean) is given by

$$\mu_1' = \tfrac{1}{2}(n + 1). \tag{2.16}$$

Similarly

$$\sum p_i^2 = \tfrac{1}{6} n(n + 1)(2n + 1)$$

and hence the variance is given by

$$\mu_2 = \frac{1}{n} \sum p_i^2 - \mu_1'^2$$

$$= \frac{1}{12}(n^2 - 1). \tag{2.17}$$

From (2.7) and (2.10) we find

$$\frac{1}{2} \sum d^2 = \frac{1}{12} n(n^2 - 1) - n \left[\frac{1}{n} \sum p_i q_i - \frac{1}{n} \sum p_i \frac{1}{n} \sum q_i \right]$$

so that the first product-moment, which is the expression in square brackets on the right, is given by

$$\mu_{11} = \frac{1}{12}(n^2 - 1) - \frac{1}{2n} \sum d^2.$$

Thus the product-moment correlation is

$$\frac{\mu_{11}}{\sqrt{\mu_2(X)\mu_2(Y)}} = 1 - \frac{6 \sum d^2}{n^3 - n} = r_s.$$

2.7 The general coefficient Γ of equation (2.1) thus embraces t, r_s and r as special cases and exhibits how we get different coefficients according to our method of scoring the differences between individuals. The scoring for t is the simplest possible and assigns a unit mark, however, near or separated the individuals are in the ranking. The scoring for r_s is more elaborate and gives greater weight to differences between individuals if they are further apart (i.e. separated by more intervening members of the ranking). The scoring for r attempts to give an objective value to the difference by measuring it on the variate scale, if one exists. The choice between these methods, or a choice of other possible methods, depends on practical considerations.

2.8 In Section 1.8 we remarked on three desirable properties for a coefficient of rank correlation. The coefficient Γ is easily seen to possess the property of varying from -1 to $+1$, because of the Cauchy–Schwartz inequality which states that

$$\left(\sum ab \right)^2 \leq \sum a^2 \sum b^2.$$

We shall now prove that under certain conditions it possesses the third property mentioned in Section 1.8: namely, if two corresponding pairs of ranks in the two rankings do not agree in order and the members of one pair are interchanged (so that they do agree) the coefficient Γ will increase, provided (1) that the scores a_{ij}, b_{ij} are not zero and (2) that the scores do not *decrease* with increasing separation of the ranks. These conditions are obeyed by t, r_s and product-moment r.

    Following our previous notation, let the rank of the rth member, p_r, be greater than that of the sth member p_s; and let the corresponding ranks in the other ranking be q_r and q_s. Let Σ'' denote summation over i and j except r and s. Then initially

$$\Sigma \, a_{ij}b_{ij} = \Sigma'' a_{ij}b_{ij} + \Sigma'' a_{rj}b_{rj} + \Sigma'' a_{ir}b_{ir} + \Sigma'' a_{sj}b_{sj}$$
$$+ \Sigma'' a_{is}b_{is} + a_{rs}b_{rs} + a_{sr}b_{sr}$$
$$= \Sigma'' a_{ij}b_{ij} + 2 \Sigma'' a_{rj}b_{rj} + 2 \Sigma'' a_{sj}b_{sj} + 2a_{rs}b_{rs}. \quad (2.18)$$

After interchanging p_r and p_s we have

$$\Sigma \, a_{ij}b_{ij} = \Sigma'' a_{ij}b_{ij} + 2 \Sigma'' a_{sj}b_{rj} + 2 \Sigma'' a_{rj}b_{sj} - 2a_{rs}b_{rs}. \quad (2.19)$$

The difference, namely the increase, is

$$-2 \Sigma'' (a_{sj} - a_{rj})(b_{sj} - b_{rj}) - 4a_{rs}b_{rs}$$
$$= -2 \Sigma (a_{sj} - a_{rj})(b_{sj} - b_{rj}). \quad (2.20)$$

Now if $p_j > p_r$ (and therefore $> p_s$) $a_{sj} > 0$, $a_{rj} \geq 0$ and, under our condition on the scores, $a_{sj} \geq a_{rj}$. If $p_r > p_j \geq p_s$, $a_{sj} \geq 0$, $a_{rj} < 0$ and hence $a_{sj} - a_{rj} = a_{sj} + a_{jr} > 0$. If $p_r > p_s > p_j$, $a_{sj} < 0$, $a_{rj} < 0$ and $a_{jr} \geq a_{js}$ so that $a_{sj} - a_{rj} \geq 0$. Thus in all cases $a_{sj} - a_{rj} \geq 0$ and in at least one case we have the strict inequality $a_{sj} - a_{rj} > 0$.

Similarly it will be found that $b_{sj} - b_{rj} \leq 0$ and is strictly negative in at least one case. Hence the summation on the right in (2.20) is negative and the increase must be positive. This completes the proof.

2.9 We shall make considerable use of the above approach in later chapters, particularly in connection with sampling problems. We now prove two of the assertions in Chapter 1, namely

(a) in Section 1.13—that the minimum number of interchanges between neighbours required to transform one ranking into another is simply related to the score S; and

(b) in Section 1.19—that the values of r_s obtained by correlating a ranking X with a ranking Y and its conjugate Y' are equal in magnitude but opposite in sign.

2.10 We will prove the second one first. It will be sufficient if we consider a ranking X typified by p_i correlated with the natural order $1, \ldots, n$ for the Y ranking and its inverse $n, \ldots, 1$ for the Y' ranking. Denoting the values of $\Sigma \, d^2$ for the second by $\Sigma' d^2$ we have

$$\Sigma \, d^2 = \Sigma (p_i - i)^2 = \Sigma p_i^2 + \Sigma i^2 - 2 \Sigma ip_i$$
$$= \tfrac{1}{3}n(n + 1)(2n + 1) - 2 \Sigma ip_i$$
$$\Sigma' d^2 = \Sigma \{p_i - (n + 1 - i)\}^2$$
$$= \tfrac{1}{3}n(n + 1)(2n + 1) - 2 \Sigma (n + 1 - i)p_i.$$

Hence

$$\sum d^2 + \sum' d^2 = \tfrac{2}{3}n(n + 1)(2n + 1) - 2 \sum (n + 1)p_i$$

$$= \tfrac{2}{3}n(n + 1)(2n + 1) - n(n + 1)^2$$

$$= \tfrac{1}{3}(n^3 - n).$$

Thus

$$r_s + r_s' = 2 - \frac{6}{n^3 - n}\{\sum d^2 + \sum' d^2\}$$

$$= 0$$

which establishes the result required.

2.11   To show that the number of interchanges s is given by

$$s = \tfrac{1}{4}n(n - 1) - \tfrac{1}{2}S \qquad (2.21)$$

we shall first prove that s is not greater than the value on the right in (2.21) and then that s cannot be less than that value. The equality will follow.

   Without loss of generality we can suppose one ranking to be in the natural order $1, 2, \ldots, n$; for we are only concerned with interchanges and can re-christen the objects so that one set is in the natural order. Let the object with rank i in the second order have rank p_i in the first, and consider the re-arrangement of the second order.

   Define an indicator function

$$m_{ij} = 1 \quad \text{if} \quad p_i > p_j$$
$$= 0 \quad \text{if} \quad p_i < p_j. \qquad (2.22)$$

The object with rank 1 may be transformed to the first place (on the extreme left) by $p_1 - 1$ interchanges. This will move the object with rank 2 to the right by m_{12} places. To transfer this to the second place will then require $p_2 - 2 + m_{12}$ interchanges. Similarly the ith object will require

$$p_i - i + m_{1i} + m_{2i} + \cdots + m_{i-1,i}$$

interchanges. Adding all these together we find for the total number of interchanges

$$\sum_{i=1}^{n} p_i - \sum_{i=1}^{n} i + \sum_{i<j} m_{ij} = \sum_{i<j} m_{ij}, \qquad (2.23)$$

where $\sum_{i<j}$ denotes summation over values for which $i < j$. Now we may write

$$S = \sum_{i<j} (1 - 2m_{ij}), \qquad (2.24)$$

for each unit contribution to S (counting only those pairs for which $i < j$ so as not to count everything twice) is $+1$ if $p_i < p_j$ and -1 in the contrary case. Thus the number of interchanges given by (2.23) is equal to

$$\tfrac{1}{2} \sum_{i<j} (1) - \tfrac{1}{2}S = \tfrac{1}{4}n(n - 1) - \tfrac{1}{2}S,$$

since the summation of unity over values for which $i < j$ is half the total number of pairs, which is $\frac{1}{2}n(n-1)$. Hence, the minimum number of interchanges s satisfies

$$s \leq \frac{1}{4}n(n-1) - \frac{1}{2}S. \tag{2.25}$$

Let T be a sequence of interchanges which reduces a given arrangement to the natural order. We classify the interchanges composing T into n groups $T_1, T_2, \ldots, T_n$. T_1 consists of those which involve the object 1, T_2 of those which involve 2 but not 1, and so on, T_n being an empty set.

In any group T_i let A_i be the number of interchanges which move the object to the right and B_i the number which move it to the left. The number of interchanges required in T_1 will be at least $p_1 - 1$. These interchanges will move object 2 m_{12} places to the right. The net result of operations in T_2 will be to move the object 2 $A_2 - B_2$ places to the right, and thus the total movement of object 2 is

$$m_{12} + A_2 - B_2,$$

and this must equal $2 - p_2$. Hence

$$B_2 - A_2 = p_2 - 2 + m_{12}$$

and hence

$$B_2 + A_2 \geq p_2 - 2 + m_{12}.$$

In a similar way we have

$$B_i + A_i \geq p_i - i + m_{1i} + \cdots + m_{i-1,i}.$$

Adding such inequalities for $i = 1, \ldots, n-1$ and remembering that

$$p_n - n + m_{1n} + \cdots + m_{n-1,n} = 0$$

we have

$$s = \sum_{i=1}^{n-1} (A_i + B_i)$$

$$\geq \sum p_i - \sum i + \sum_{i<j} m_{ij}$$

$$\geq \frac{1}{4}n(n-1) - \frac{1}{2}S \tag{2.26}$$

and thus from (2.25) and (2.26) we establish (2.21).

2.12 We will now prove and develop the result of equations (1.12) and (1.13), exhibiting r_s as a coefficient of weighted inversions. We will suppose, without loss of generality, that one ranking is in the natural order $1, 2, \ldots, n$. Defining m_{ij} as in (2.22) we have

$$V = \sum_{i<j} m_{ij}(j - i). \tag{2.27}$$

Adding and subtracting $\sum_{i>j} j m_{ij}$ we have

$$V = \sum_{i<j} j m_{ij} + \sum_{i>j} j m_{ij} - \sum_{i<j} i m_{ij} - \sum_{i>j} j m_{ij}$$

$$= \sum_{i,j} j m_{ij} - \sum_{i<j} i(m_{ij} + m_{ji}).$$

Here p_j is the rank in the first ranking corresponding to j in the second; we sum over i in the first summation and j in the second to find

$$V = \sum_{j} j(n - p_j) - \sum_{i} i(n - i)$$

$$= \sum_{i} i^2 - \sum_{i} i p_i. \tag{2.28}$$

As in (2.7) we have

$$\sum d^2 = 2 \sum i^2 - 2 \sum i p_i,$$

and hence

$$V = \tfrac{1}{2} \sum d^2. \tag{2.29}$$

It follows at once that

$$r_s = 1 - \frac{12V}{n^3 - n}. \tag{2.30}$$

2.13 Another form for r_s which is useful in certain contexts, though not as a convenient form for computation, is as follows: let a_{ij}, b_{ij} be the unit scores entering into the expression for Γ. Put

$$a_i = \sum_{j=1}^{n} a_{ij}$$

$$b_i = \sum_{j=1}^{n} b_{ij}.$$

Then

$$r_s = \frac{\sum_{i=1}^{n} a_i b_i}{\tfrac{1}{3}(n^3 - n)}. \tag{2.31}$$

In fact, our scores a_{ij} may be expressed in terms of m_{ij} by the relation

$$a_{ij} = 1 - 2 m_{ij},$$

for this is $+1$ if $p_i < p_j$ and -1 if $p_i > p_j$. Furthermore the rank of any member may be expressed as

$$p_i = \sum_{j=1}^{n} m_{ij} + 1. \tag{2.32}$$

Hence, we have

$$p_i = -\tfrac{1}{2} \sum_{j} a_{ij} + \tfrac{1}{2}(n + 1),$$

and, remembering that the mean of the p's, $\bar{p}$, is $\frac{1}{2}(n + 1)$, we find

$$a_i = -2(p_i - \bar{p}). \tag{2.33}$$

Thus, since r_s is the product-moment correlation of p and q (Section 2.4) we have

$$r_s = \frac{\frac{1}{4}\sum_i a_i b_i}{\frac{1}{12}(n^3 - n)}$$

$$= \frac{\sum a_i b_i}{\frac{1}{3}(n^3 - n)},$$

as stated in (2.31).

2.14 It follows that we can express r_s in the form

$$r_s = \frac{3}{n^3 - n} \sum a_{ij} b_{ik} \tag{2.34}$$

where the summation extends over all i, j, k. Or again, if we consider summation for which $j \neq k$,

$$r_s = \frac{3}{n^3 - n} \left(\sum a_{ij} b_{ij} + \sum_{j \neq k} a_{ij} b_{ik} \right)$$

$$= \frac{3}{(n + 1)} t + \frac{3}{n^3 - n} \sum_{j \neq k} a_{ij} b_{ik}. \tag{2.35}$$

It is interesting to consider this form in terms of agreements (with respect to order) of pairs from two rankings. If a_{ij} and b_{ij} are of the same sign we may say that there is a *concordance of type 1*. (We are dealing only with the case of unitary scores.) If a_{ij} and b_{ik} are of the same sign we may say that there is a *concordance of type 2*. If the signs are opposite we may speak of *discordances*. Then t is simply related to the proportion of concordances of type 1 in the two rankings. The number of possible concordances of type 2 is $n(n - 1)(n - 2)$, and if we denote this by k_2 to distinguish it from the number of concordances of type 1 (equivalent to $2P$ of equation (1.6)), which we will call k_1, we have

$$\sum a_{ij} b_{ij} = k_1 - \{n(n - 1) - k_1\}$$

$$= 2k_1 - n(n - 1),$$

$$\sum a_{ij} b_{ik} = 2k_2 - n(n - 1)(n - 2),$$

and on substitution in (2.35) we find

$$r_s = \frac{6(k_1 + k_2)}{n^3 - n} - \frac{3(n - 1)}{n + 1}. \tag{2.36}$$

Proof of Daniels' inequality

2.15   Since relations like $p_i > p_j > p_k > p_i$ are impossible, a_{ij}, a_{jk}, a_{ki} cannot all have the same sign and hence their sum must be ± 1. Similarly for the b-scores. Put

$$(a_{ij} + a_{jk} + a_{ki})(b_{ij} + b_{jk} + b_{ki}) = \varepsilon_{ijk}. \tag{2.37}$$

Now let the corresponding ranks p_i, p_j, p_k, q_i, q_j, q_k be renamed according to their order of magnitude, becoming p_i', p_j', etc. For example, a set of p-ranks 472 becomes 231. Then ε_{ijk} is $+1$ or -1 according as p' is an even or odd permutation of q', that is to say if it takes an even or odd number of inter-changes of consecutive members to go from one to the other. If any pair of i, j, k are equal ε is zero.

Summing over all values of the suffixes, we find

$$3n \sum a_{ij} b_{ij} - 6 \sum a_{ij} b_{ik} = \sum \varepsilon_{ijk}. \tag{2.38}$$

Now there are $n(n - 1)(n - 2)$ possible triplets of (i, j, k) and $\sum \varepsilon_{ijk}$ is the number of them which are 'even'. Put

$$U = \sum \varepsilon_{ijk} / n(n - 1)(n - 2). \tag{2.39}$$

We have $-1 \le U \le 1$ and hence, from (2.38)

$$-1 \le \frac{3(n + 2)}{n - 2} t - \frac{2(n + 1)}{n - 2} r_s \le 1. \tag{2.40}$$

In the limit for n large this becomes

$$-1 \le 3t - 2r_s \le 1. \tag{2.41}$$

That the upper limit is attainable for $t \ge 0$ is seen from consideration of a ranking in which the q's are the natural order $1, \ldots, n$ and the p's are a cyclic permutation of the natural order, say $m + 1, m + 2, \ldots, n, 1, 2, \ldots, m$. All the εs are then $+1$ and U attains its maximum.

Proof of the Durbin–Stuart inequality

2.16   The problem of finding an attainable upper limit of r_s for given $t \ge 0$ is that of finding the smallest value of V for given Q. Consider, for example, a population of six with $t = 0.6$ and hence $Q = 3$. If the first ranking is in the natural order the value of V is least when the second ranking is 214365, for we have constructed three inversions 21, 43, 65 for which the weight has its minimum value unity.

Let us then define a *compact* set of ranks as one whose members form inversions only with each other; no group being subdivisible into subgroups of the same type. All ranks to the left of a compact set are less than every one of its members; those to the right are greater.

Consider any inversion of weight 2, say $r + 2, r$. There are three possible orders of the triad $r, r + 1, r + 2$, namely $r + 2, r + 1, r$; $r + 2, r, r + 1$ and $r + 1, r + 2, r$. In a minimal ranking, i.e. a ranking with minimal V for given Q, all triads spanned by inversions of weight two must be of type $r + 2$, $r + 1, r$, except possibly for the triads of a single compact set. For suppose that in addition to the triad $r + 2, r, r + 1$ there is a further non-overlapping triad $x + 2, x, x + 1$. On replacing these by $r + 2, r + 1, r$ and $x + 1, x, x + 2$ we reduce V but leave Q unchanged.

There are twelve possible orders of a tetrad spanned by the inversion $r + 3, r$ of weight three. Of these the only one which does not contain a triad of types $r + 2, r, r + 1$ or $r + 1, r + 2, r$ is $r + 3, r + 2, r + 1, r$. Hence in the minimal ranking all tetrads are of this type, except possibly for one compact set.

By arguing in this way we see that the minimal ranking must consist of sets of the form $q + r, q + r - 1, \ldots, r$, except possibly for a single compact set which we shall call the residual set. It is this residual which causes most of the trouble in proving the desired result. In particular it may comprise the whole ranking.

2.17 Taking first the case when there is no residual set, suppose there are α_1 sets of form $r + q_1, \ldots, r$; α_2 sets of form $r + q_2, \ldots, r$; and so on. Then

$$\sum \alpha_i q_i = n \tag{2.42}$$

$$Q = \tfrac{1}{2} \sum \alpha_i q_i (q_i - 1)$$
$$= \tfrac{1}{2} (\sum \alpha_i q_i^2 - n) \tag{2.43}$$

$$V = \tfrac{1}{6} \sum \alpha_i q_i (q_i^2 - 1)$$
$$= \tfrac{1}{6} (\sum \alpha_i q_i^3 - n). \tag{2.44}$$

Now by the Cauchy-Schwartz inequality

$$(\sum \alpha_i q_i)(\sum \alpha_i q_i^3) \geq (\sum \alpha_i q_i^2)^2$$

and on substitution from (2.42)–(2.44) we find

$$V \geq \tfrac{2}{3} Q \left(1 + \frac{Q}{n}\right). \tag{2.45}$$

The equality holds when all qs are equal, as is possible in certain cases.

2.18 We must now deal with the case when there is a residual set S of s ranks which is not of the form $I = r + s, r + s - 1, \ldots, r$. We suppose that S has q inversions of total weight v and that I has k more inversions than has S. We may thus proceed from I to S by k interchanges of neighbouring pairs. If we choose such a sequence which gives the greatest reduction in v we arrive at a ranking with q inversions and minimum v, which is what we are seeking.

If we start from I the maximum reduction for k interchanges ($k \le s - 1$) is $\frac{1}{2}k(k + 1)$. The sequence of interchanges which gives this reduction is obtained by moving the smallest rank from the extreme right-hand of I towards the left. When the left-hand end is reached, the requirement that S should be a compact set is violated, since the smallest rank then itself forms a single-member compact set.

Further, whenever, starting from I, any sequence of $k \ge s - 1$ interchanges is carried out, the final result cannot be a residual set forming part of a minimal ranking, for it is always possible, by moving the smallest rank of I through $(s - 1)$ interchanges as above, to violate the condition of compactness while producing the same number q of inversions with a smaller v.

Thus, for a residual set S of s ranks forming part of a minimal ranking, only one rank will be displaced from its position in I, and the number of interchanges transforming I to S will be $k \le s - 2$.

2.19   We now show that the inequality (2.45) is satisfied by the residual set alone. In fact the inequality holds strictly, i.e.

$$v > \frac{2}{3}q\left(1 + \frac{q}{s}\right) = \frac{2}{3}\frac{q}{s}(q + s).$$

This is true if

$$\tfrac{1}{6}s(s^2 - 1) - \tfrac{1}{2}k(k + 1) > \tfrac{2}{3}s\{\tfrac{1}{2}s(s - 1) - k\}\{\tfrac{1}{2}s(s - 1) + s - k\}$$

$$> \tfrac{1}{6}s\{s^2(s^2 - 1) - 4s^2k + 4k^2\}$$

i.e. if

$$k < \frac{s(4s - 3)}{3s + 4}. \tag{2.46}$$

Now $k \le s - 2$, so (2.46) is certainly true if

$$s - 2 < \frac{s(4s - 3)}{3s + 4},$$

i.e. if $s^2 - s + 8 > 0$, which is true. Thus the inequality (2.45) holds strictly for the residual set alone.

Let Q' be the number of inversions in the remainder of the ranking apart from the residual set, and let V' be the corresponding weighted sum.

Then

$$Q = Q' + q$$

and

$$V = V' + v$$

as a consequence of the definition of compact sets.

Also, from (2.45)

$$V' \ge \frac{2}{3}\left(\frac{Q'^2}{n - s} + Q'\right)$$

and

$$v > \frac{2}{3}\left(\frac{q^2}{s} + q\right).$$

We need to show that (2.45) is generally true, i.e. that

$$V' + v \geq \frac{2}{3}\left[\frac{(Q' + q)^2}{n} + Q' + q\right].$$

This is satisfied if

$$\frac{Q'^2}{n - s} + \frac{q^2}{s} \geq \frac{(Q' + q)^2}{n},$$

i.e. if

$$n\{sQ'^2 + (n - s)q^2\} \geq s(n - s)(Q' + q)^2$$

which is satisfied if

$$s^2Q'^2 + (n - s)^2q^2 - 2s(n - s)Q'q \geq 0,$$

i.e. if

$$[sQ' - (n - s)q]^2 \geq 0,$$

which is true. Thus (2.45) is established. The equality is only attainable when there is no residual set.

Equation (1.21) follows. To find the corresponding expression for negative *t* we replace the ranking concerned by its conjugate. This merely changes the sign of r_s and *t*.

Spearman's footrule

2.20 To conclude this chapter we will consider briefly a coefficient based, not on $\sum d^2$ like r_s, but on $\sum |d|$ where $|d|$ stands for the absolute value of *d*, i.e. its value without regard to sign. This coefficient, sometimes known as Spearman's 'footrule', is not of the general type of (2.1). Let us put

$$R = 1 - \frac{3 \sum |d|}{n^2 - 1}. \tag{2.47}$$

If two rankings are identical, $\sum |d| = 0$ and $R = 1$ as we should require.

Now if one ranking is the inverse of the other, suppose that *n* is odd and equals $2m + 1$. Then, as in Section 1.15,

$$\sum |d| = 2\{2m + (2m - 2) + (2m - 4) + \cdots + 4 + 2\}$$

$$= 2m(m + 1).$$

Hence

$$R = 1 - \frac{6m(m + 1)}{(2m + 1)^2 - 1}$$

$$= -0.5. \tag{2.48}$$

If n is even, say $2m$, we find similarly

$$\Sigma |d| = 2m^2,$$

and hence

$$R = 1 - \frac{6m^2}{4m^2 - 1}$$

$$= -0.5\left[1 + \frac{3}{n^2 - 1}\right]. \tag{2.49}$$

Thus the coefficient R cannot have a minimum value -1 unless $n = 2$. For large even n it rapidly approaches -0.5 and for odd n it must be -0.5 in all cases. This is a defect which cannot be entirely remedied by taking a different multiplier from $3/(n^2 - 1)$ for $\Sigma |d|$ in (2.47).

If we were to write

$$R' = 1 - \frac{4 \Sigma |d|}{n^2},$$

R' would become -1 for even n and $-1 + 2/n^2$ for odd n.

2.21   Moreover, R is much less sensitive than t or r_s. If, for example, we write down the 24 permutations of 1 to 4 and correlate them with the natural order we find:

Values of r_s	Values of R	Frequencies
1.0	1.0	1
0.8	0.6	3
0.6	0.2 ⎫	1
0.4	0.2 ⎬	4
0.2	0.2 ⎭	2
0.0	−0.2 ⎫	2
−0.2	−0.2 ⎪	2
−0.4	−0.2 ⎬	4
−0.6	−0.2 ⎭	1
−0.8	−0.6 ⎫	3
−1.0	−0.6 ⎭	1
		⎯⎯
		24

For the same value of R, e.g. 0.2, r_s may vary from 0.2 to 0.6; and for $R = -0.2$ r_s may vary from 0.0 to -0.6.

These features make tau and rho preferable to Spearman's footrule R. Nevertheless, interest in R has increased in recent years.

Notes and references

The main early paper on generalised coefficients was Daniels (1944). See also the references to Chapter 1. Hoeffding (1948a) considered a very general class of statistics, now known as U-statistics, which include rank correlation coefficients as a special case.

See Daniels (1948) for the property of Section 2.8. See the references to Chapter 1 for the inequalities on r_s and t.

The original reference on Spearman's footrule is Spearman (1906). Diaconis and Graham (1977) studied its properties along with those of rho and tau. Ury and Kleinecke (1979) gave tables of the exact null distribution for small samples and Franklin (1988b) extended them up to $n = 18$.

Farlie (1960, 1961, 1963) considered the asymptotic relative efficiency of Daniels' generalised correlation coefficients. Blum, Kiefer and Rosenblatt (1961) studied the asymptotic behaviour of U-statistics.

Chapter 3

Tied Ranks

3.1 In practical applications of ranking methods there sometimes arise cases in which two or more individuals are so similar that no preference can be expressed between them. When an observer is ranking members by subjective judgments this effect may be due either to a genuine indistinguishability of the objects or to a failure by the observer to distinguish any differences that exist. The members are then said to be *tied*. The arrangement of students in order of merit or by reference to examination marks is a familiar source of ties of this kind.

3.2 The method which we shall adopt for allocating rank-numbers to tied individuals is to average the ranks which they would possess if they were not tied. For instance, if the third and fourth members are tied, each is allotted the number $3\frac{1}{2}$, and if the second to the seventh inclusive are tied, each is allotted the number $\frac{1}{6}(2 + 3 + 4 + 5 + 6 + 7) = 4\frac{1}{2}$. This is known as the 'midrank method'. When there is nothing to choose between individuals, we must clearly rank them all alike if we rank them at all; and our method has the advantage that the sum of the ranks for all members remains the same as for an untied ranking. In general, if ties occur for the ith to kth inclusive members, the midrank is $(i + k)/2$.

We have now to consider the effect of ties on the calculation of t and r_s.

Calculation of t

3.3 In Section 1.9 we saw that a score of $+1$ or -1 was allotted to a pair of members according as their ranks were in the natural order or not. If they are tied we shall allot the score zero, midway between the two values which they would have if they were not tied. The score S is then easily calculated as $P - Q$.

3.4 A new point arises, however, in regard to the calculation of the denominator by which S is to be divided to obtain t. We have two possibilities:

(a) We can use the denominator $\frac{1}{2}n(n - 1)$ as for the untied form of t; we note however that t cannot be calculated from either (1.5) or (1.6) when there are ties because $P + Q \neq n(n - 1)/2$ when one or more pairs have a score of zero.

(b) We can replace $\frac{1}{2}n(n - 1)$ by $\frac{1}{2}\sqrt{(\sum a_{ij}^2 \sum b_{ij}^2)}$ where a_{ij} is the score of the ith and jth members in one ranking and b_{ij} is the corresponding score in the other.

Where no ties exist, any term a_{ij}^2 is unity, so that $\sum a_{ij}^2$ reduces to the number of possible terms, namely $n(n - 1)$; similarly for $\sum b_{ij}^2$ so that the expression $\frac{1}{2}\sqrt{(\sum a_{ij}^2 \sum b_{ij}^2)}$ reduces to $\frac{1}{2}n(n - 1)$, as it should. The reason for adopting this expression in the tied case is clear from Sections 2.2 and 2.3.

If there is a tie of u consecutive members all the scores arising from any pair chosen from them is zero. There are $u(u - 1)$ such pairs. Consequently the sum $\sum a_{ij}^2$ is $n(n - 1) - \sum u(u - 1)$, where $\sum$ the summation is over all sets of u tied scores. If we define

$$U = \tfrac{1}{2} \sum u(u - 1) \qquad (3.1)$$

for ties in one ranking and

$$V = \tfrac{1}{2} \sum v(v - 1) \qquad (3.2)$$

for ties in the other, our alternative form of the coefficient t for tied ranks may be written

$$t_b = \frac{S}{\sqrt{[\tfrac{1}{2}n(n - 1) - U]}\sqrt{[\tfrac{1}{2}n(n - 1) - V]}}. \qquad (3.3)$$

Before discussing the alternative forms further, let us consider a numerical example.

Example 3.1

Two rankings are given as follows:

X	1	$2\frac{1}{2}$	$2\frac{1}{2}$	$4\frac{1}{2}$	$4\frac{1}{2}$	$6\frac{1}{2}$	$6\frac{1}{2}$	8	$9\frac{1}{2}$	$9\frac{1}{2}$
Y	1	2	$4\frac{1}{2}$	$4\frac{1}{2}$	$4\frac{1}{2}$	$4\frac{1}{2}$	8	8	8	10

Except for ties, both rankings are in the same order and the correlation is high. Considering the first member in association with the other 9, we see that the contribution to S is 9; the second and third members of X are tied, so that they contribute nothing to S, whatever the Y-score for this pair. The score of members associated with the second member will be found to be 7, and so on. The full score is

$$9 + 7 + 4 + 4 + 4 + 3 + 1 + 1 + 0 = 33.$$

If we adopt alternative (*a*) of Section 3.4 the value of t is then given by

$$t_a = \frac{33}{45} = +0.733.$$

Under alternative (b) we have, for the X-ranking,

$$U = [2(1) + 2(1) + 2(1) + 2(1)]/2 = 4$$

and for the Y-ranking

$$V = [4(3) + 3(2)]/2 = 9.$$

Hence, from (3.3)

$$t_b = \frac{33}{\sqrt{[(41)(36)]}}$$

$$= +0.859.$$

The value given by alternative (b) must, of course, be greater than that of (a) in all cases. In the present instance it is substantially greater.

3.5   From the general point of view developed in Chapter 2, the appropriate form of coefficient, as a true measure of correlation between two sets of numbers, is t_b. For example, if we are measuring the agreement between two judges in arranging a set of candidates in order of merit (no objective order necessarily existing) we should use t_b. Both judges may be wrong in relation to some objective order, and they may disagree with other judges, but that is not the point. We are measuring their agreement, not their accuracy.

Suppose, for instance, that both rankings are the same, that the last member of each is n, and that all the others are tied and hence have average rank $\frac{1}{2}n$. Then $t_b = 1$, as it should be to express complete agreement between the rankings. But we also have, from (3.1) and (3.2),

$$U = V = \tfrac{1}{2}(n - 1)(n - 2)$$

and hence

$$\tfrac{1}{2}n(n - 1) - U = n - 1.$$

Thus, since the score S is also $n - 1$ (confirming that $t_b = 1$), we have

$$t_a = \frac{n - 1}{\tfrac{1}{2}n(n - 1)} = \frac{2}{n} \tag{3.4}$$

and for large n this is nearly zero. Clearly t_a is an inappropriate measure of *agreement*.

3.6   Nevertheless, there may be cases where t_a is a better measure than t_b. Suppose that there really eixsts an objective order. The purpose of correlating a ranking assigned by an observer is then to measure the accuracy. The form t_b would give weight to the fact that if the observer produces ties the variation in the estimates is reduced. The calculation takes into account, so to speak, the clustering of values *in spite of the fact that there should be no ties because there really is an objective order.* In such a case it may be argued that the full divisor $\tfrac{1}{2}n(n - 1)$ should be used in calculating t, i.e. that t_a is the appropriate form.

Consider the case where our ranking is the natural order $1, \ldots, n$ and the other has the first $(n - 1)$ members tied (with ranks each $\tfrac{1}{2}n$) and the last member ranked as n. As in (3.4) we have

$$t_a = \frac{2}{n},$$

whereas

$$t_b = \frac{n - 1}{\sqrt{[\frac{1}{2}n(n - 1)^2]}} = \sqrt{\frac{2}{n}}. \tag{3.5}$$

For example, with $n = 9$, $t_a = 0.22$ and $t_b = 0.47$. The first value seems nearer to what we should expect of a measure of agreement with an objective order. The observer has not put any pair in the wrong order, and has ranked one member correctly; but the observer has been unable to distinguish between the first nine, and a value of 0.22 as a measure of ability seems appropriate. On the other hand, if the first ranking were not known to be objective but was just an expression of opinion from another observer of no greater known reliability, a value of 0.47 seems a fair measure of concordance.

3.7 There is another interesting way of looking at the problem of tied ranks. Suppose we regard any tied set u as due to inability to distinguish real differences. We may then ask: what is the *average* value of t over all the $u!$ possible ways of assigning integer ranks to the tied members?

   If we replace any tied set u by integer ranks and average for all $u!$ possible orders we get the same result as by replacing the scores a_{ij} for the tied members by zero; for in the $u!$ arrangements each pair will occur an equal number of times in the order AB and in the order BA, so that the allocation of $+1$ in the one case and -1 in the other is equivalent to allocating zero on the average. Thus we may regard t_a as an average coefficient, such as would be obtained if the tied ranks were replaced by integer ranks in all possible ways, t calculated for each, and an arithmetic mean taken of the resulting values.

Calculation of rho

3.8 We turn to consider the analogous problems for the rank correlation coefficient r_s. Again we shall have the choice of two denominators and two coefficients, which we may denote by r_{sa} and r_{sb}. If there are sets of ties in the two rankings typified by u and v we define

$$U' = \tfrac{1}{12} \sum (u^3 - u)$$
$$V' = \tfrac{1}{12} \sum (v^3 - v) \tag{3.6}$$

Then we have

$$r_{sa} = 1 - \frac{6 \sum d^2 + U' + V'}{n^3 - n} \tag{3.7}$$

$$r_{sb} = \frac{\tfrac{1}{6}(n^3 - n) - \sum d^2 - U' - V'}{\sqrt{[\frac{1}{6}(n^3 - n) - 2U'][\frac{1}{6}(n^3 - n) - 2V']}}. \tag{3.8}$$

We shall prove these formulae below, but before doing so we will consider an example.

Example 3.2

Consider again the two rankings of Example 3.1:

| X | 1 | $2\frac{1}{2}$ | $2\frac{1}{2}$ | $4\frac{1}{2}$ | $4\frac{1}{2}$ | $6\frac{1}{2}$ | $6\frac{1}{2}$ | 8 | $9\frac{1}{2}$ | $9\frac{1}{2}$ |
| Y | 1 | 2 | $4\frac{1}{2}$ | $4\frac{1}{2}$ | $4\frac{1}{2}$ | $4\frac{1}{2}$ | 8 | 8 | 8 | 10 |

In the first ranking there are four tied pairs ($u = 2$) and hence

$$U' = \tfrac{4}{12}(2^3 - 2) = 2.$$

In the second there is one set for which $v = 4$ and one for which $v = 3$, so that

$$V' = \tfrac{1}{12}(4^3 - 4 + 3^3 - 3)$$

$$= 7.$$

We also find

$$\Sigma d^2 = 13.$$

Hence, from (3.7),

$$r_{sa} = 1 - \frac{6(13 + 7 + 2)}{990}$$

$$= 0.867$$

and from (3.8)

$$r_{sb} = \frac{165 - 22}{\sqrt{[(161)(151)]}}$$

$$= 0.917.$$

3.9   It is useful to note that (3.8) can be put in the form

$$r_{sb} = \frac{\tfrac{1}{6}(n^3 - n) - \Sigma d^2 - U' - V'}{\{\tfrac{1}{6}(n^3 - n) - (U' + V')\}\sqrt{\left[1 - \frac{(U' - V')^2}{\{\tfrac{1}{6}(n^3 - n) - (U' + V')\}^2}\right]}}. \tag{3.9}$$

Thus, if U' and V' are small compared with $\tfrac{1}{6}(n^3 - n)$, we have approximately (and exactly if $U' = V'$)

$$r_{sb} = 1 - \frac{\Sigma d^2}{\tfrac{1}{6}(n^3 - n) - (U' + V')} \tag{3.10}$$

or, slightly more approximately,

$$r_{sb} = 1 - \frac{\Sigma d^2}{\tfrac{1}{6}(n^3 - n)}, \tag{3.11}$$

which is the ordinary formula for rho in the untied case. We therefore expect that when the ties are not very numerous the use of the formulae (3.9) or (3.10) will make little difference to the numerical values given by the use of (3.11).

For instance, in the data of Example 3.2 we find for the form (3.10)

$$r_{sb} = 1 - \frac{13}{165 - 9}$$

$$= 0.9167$$

and for the form (3.11)

$$r_{sb} = 1 - \frac{13}{165} = 0.9212.$$

The value given by (3.8) is 0.9171. All three values agree to the nearest second decimal place.

3.10 To establish formulae (3.7) and (3.8) we have to use some of the results of Chapter 2. We saw in Section 2.6 that rho may be regarded as the product-moment correlation between the ranks. Suppose we adopt the same viewpoint when some of the ranks are tied.

For a set of untied ranks $p_1, p_2, \ldots, p_n$, the sum of squares of ranks is

$$\Sigma \, p_i^2 = \tfrac{1}{6}n(n + 1)(2n + 1)$$

and the sum of ranks is

$$\Sigma \, p_i = \tfrac{1}{2}n(n + 1).$$

If a set of u ranks are tied the sum of ranks remains the same but the sum of squares is altered. Suppose the ranks $p_k + 1, \ldots, p_k + u$ are tied. Then the sum of squares is reduced by

$$(p_k + 1)^2 + (p_k + 2)^2 + \cdots + (p_k + u)^2 - u\{p_k + \tfrac{1}{2}(u + 1)\}^2$$

$$= up_k^2 + 2p_k(1 + 2 + \cdots + u) + 1^2 + 2^2 + \cdots + u^2$$

$$- u\{p_k^2 + p_k(u + 1) + \tfrac{1}{4}(u + 1)^2\}$$

$$= \tfrac{1}{12}(u^3 - u).$$

For an untied ranking the variance is

$$\frac{1}{n} \Sigma \, \{p_i - \tfrac{1}{2}(n + 1)\}^2 = \frac{1}{n} \Sigma \, p_i^2 - \tfrac{1}{4}(n + 1)^2 = \frac{1}{12}(n^2 - 1).$$

It follows that for a tied ranking the variance is

$$\frac{1}{12}(n^2 - 1) - \frac{1}{12n} \Sigma \, (u^3 - u) = \frac{1}{2n}\{\tfrac{1}{6}(n^3 - n) - 2U'\}. \qquad (3.12)$$

Since r_{sb} may be defined as

$$r_{sb} = \frac{\text{cov}(p, q)}{\sqrt{(\text{var } p \text{ var } q)}}$$

and

$$\text{var } p + \text{var } q - 2 \text{ cov}(p, q) = \text{var}(p - q),$$

we can write

$$r_{sb} = \frac{1}{2} \frac{\text{var } p + \text{var } q - \text{var}(p - q)}{\sqrt{(\text{var } p \text{ var } q)}}. \tag{3.13}$$

It is easily verified that this gives r_{sb} for the untied case. Let us apply it to the tied case. Since $\text{var}(p - q)$ is still equal to $(1/n) \sum d^2$ we find from (3.13), on using (3.12),

$$r_{sb} = \frac{1}{2} \frac{\frac{1}{6}(n^3 - n) - 2U' + \frac{1}{6}(n^3 - n) - 2V' - 2 \sum d^2}{\sqrt{[[\frac{1}{6}(n^3 - n) - 2U'][\frac{1}{6}(n^3 - n) - 2V']]}}$$

which reduces to (3.8). The formula of (3.7) for r_{sa} follows in a similar way.

3.11   The earlier statements about the different circumstances in which t_a and t_b may be preferred applies also to a choice between r_{sa} and r_{sb}. To complete the analogy we need only prove that r_{sa} is the average of the values of coefficients which would be obtained if the ties were replaced by integer ranks in all possible ways.

If the X-ranking is held fixed, then the average covariance for all $u!$ arrangements of a set of u ranks in the other ranking Y is the covariance of the fixed ranks in X and the average of the ranks in Y; but this latter provides the values of the tied ranks. It follows that the average covariance is the covariance of the tied rankings, for the effects of different sets of ties are additive; and thus the result follows.

Example 3.3

If two rankings are identical, the last member in each is ranked n, and the other $n - 1$ are tied with average rank $\frac{1}{2}n$, we clearly have

$$r_{sb} = 1.$$

But for r_{sa} we find

$$U' = V' = \tfrac{1}{12}\{(n - 1)^3 - (n - 1)\}$$

$$= \tfrac{1}{12}n(n - 1)(n - 2)$$

and $\sum d^2 = 0$.

Thus, from (3.7),

$$r_{sa} = \frac{\frac{1}{6}(n + 1)(n)(n - 1) - \frac{1}{6}n(n - 1)(n - 2)}{\frac{1}{6}(n + 1)(n)(n - 1)}$$

$$= \frac{3}{n + 1}.$$

We find the same kind of difference between the two types of rho as between the two types of tau in Section 3.5.

3.12 A fairly common problem in psychology is to measure the relationship between two qualities, one of which provides a ranking and the other a *dichotomy* or classification into two classes according as the individual possesses a certain attribute or not. Consider the following ranking of 15 girls and boys according to merit in an examination:

Rank:	1	2	3	4	5	6	7	8	9	10	11	12	13	14	15
Sex:	B	B	G	B	G	G	B	B	B	G	B	G	B	G	G

We are interested here in whether there is any connection between sex and success—whether the boys did better than the girls on the average or vice versa.

We will imagine that the division into sex is itself a ranking. There are 8 boys and 7 girls and we will suppose that the first 8 members of the ranking by sex are tied, and also for the next group of 7. The average of the tied ranks will be $4\frac{1}{2}$ in the first case and 12 in the second, so that the pair of rankings may be written:

Ranking X:	1	2	3	4	5	6	7	8	9	10	11	12	13	14	15
Ranking Y:	$4\frac{1}{2}$	$4\frac{1}{2}$	12	$4\frac{1}{2}$	12	12	$4\frac{1}{2}$	$4\frac{1}{2}$	$4\frac{1}{2}$	12	$4\frac{1}{2}$	12	$4\frac{1}{2}$	12	12

We may now calculate t_b for these rankings. We find

$$S = 7 + 7 - 6 + 6 - 5 - 5 + 4 + 4 + 4 - 2 + 3 - 1 + 2 + 0$$

$$= 18$$

$$U = 0$$

$$V = \tfrac{1}{2}(8)(7) + \tfrac{1}{2}(7)(6) = 49$$

$$t_b = \frac{18}{\sqrt{(105)(56)}} = +0.24$$

This indicates some positive correlation between order of success and the order of sex, which we have chosen by putting boys first. (Had we put girls first, of course, we should have obtained $t_b = -0.24$ leading to the same conclusion.) In short, the boys seem to be somewhat better than the girls. Whether the evidence is sufficiently indicative of 'real' correlation is a matter of significance which we shall discuss in the next chapter.

The S and therefore the tau coefficients in the examples given in this section are linear functions of the well-known Wilcoxon sum of ranks and Mann–Whitney U tests for two independent samples. In the example concerning merit scores of boys and girls, the girls and boys are the two independent samples. The number of concordant pairs is equivalent to the number of times a G follows a B in the sex ranking or $7 + 7 + 6 + 4 + 4 + 4 + 3 + 2 = 37$ and the number of discordant pairs is the number of times a B follows a G or $6 + 5 + 5 + 2 + 1 = 19$, giving $S = 37 - 19 = 18$ as before.

Example 3.4

Seventeen male factory workers aged 50–59 years were interviewed and assessed as either efficient or overactive. These same men were also ranked according to frequency of errors on the job (least errors given lowest rank). The data are as follows:

Efficient: $2\frac{1}{2}$ $2\frac{1}{2}$ $2\frac{1}{2}$ $2\frac{1}{2}$ $6\frac{1}{2}$ $6\frac{1}{2}$ 10   10   10   10   14   14
Overactive: 5   10   14   16   17

A cursory inspection of the figures indicates that the overactive group had the greatest frequency of errors. Let us measure the relationship with t_b. We find, writing E and O for efficient and overactive respectively:

Ranking X: $2\frac{1}{2}$ $2\frac{1}{2}$ $2\frac{1}{2}$ $2\frac{1}{2}$ 5 $6\frac{1}{2}$ $6\frac{1}{2}$ 10 10 10 10 10 14 14 14 16 17
Ranking Y: E E E E O E E O E E E E E E O O O

If we wish we can replace the Es and Os by rankings, but this is unnecessary for the calculation of S. We find

$$S = 5 + 5 + 5 + 5 - 8 + 4 + 4 - 2 + 3 + 3 + 3 + 3 + 2 + 2$$

$$= 34.$$

$$U = \tfrac{1}{2}(4)(3) + \tfrac{1}{2}(2)(1) + \tfrac{1}{2}(5)(4) + \tfrac{1}{2}(3)(2)$$

$$= 20.$$

$$V = \tfrac{1}{2}(12)(11) + \tfrac{1}{2}(5)(4)$$

$$= 76.$$

Thus

$$t_b = \frac{34}{\sqrt{[(116)(60)]}} = +0.41.$$

The reader should check the calculation of S. Taking the first member in ranking X, for example, we see that no contribution to S can arise from the other members in X with rank $2\frac{1}{2}$, so that it is only subsequent members that need be considered. The first member of ranking Y, being an E, contributes $+1$ only with members which are O, of which there are five in the subsequent members.

Similarly for the second, third and fourth members of X. When we come to the fifth, with rank 5, all subsequent members of X are liable to contribute. In the Y-ranking the fifth member is an O, and hence contributions of -1 arise from subsequent members which are E, 8 in number. And so on.

3.13   In general, if a ranking consists of a dichotomy with x and $n - x = y$ members in the two classes

$$\tfrac{1}{2}n(n - 1) - U = \tfrac{1}{2}n(n - 1) - \tfrac{1}{2}x(x - 1) - \tfrac{1}{2}(n - x)(n - x - 1)$$

$$= xy.$$

Thus we have

$$t_b = \frac{S}{\sqrt{\{xy[\frac{1}{2}n(n-1) - V]\}}}.$$ (3.14)

3.14 Consider now the extreme case when both rankings consist of a dichotomy, one, say, into x and $n - x = y$ members, the other into p and $n - p = q$ members. We then have for the denominator in t_b the simple expression $\sqrt{(xypq)}$. We may express the data in what is usually known as a 2×2 contingency table, as follows:

Second quality	First quality		
	Possessing	Not-possessing	Totals
Possessing	a	b	p
Not-possessing	c	d	q
Totals	x	y	n

(3.15)

Any member of the class possessing both qualities (a in number) taken with any member of the class not possessing either (d in number) gives a pair with the same order in either ranking and hence contributes $+1$ to S. Similarly any member of the b-class with any member of the c-class contributes -1. The others contribute nothing. Hence

and

$$S = ad - bc$$

$$t_b = \frac{ad - bc}{\sqrt{(xypq)}}.$$ (3.16)

This is one useful form of a coefficient measuring association in a 2×2 table. There are other coefficients of the same kind, but it is interesting that for the extreme case when both rankings are so tied as to be dichotomous the coefficient tau is related to the usual chi square statistic calculated for 2×2 contingency tables. The relationship is $t_b^2 = \chi^2/n$.

Example 3.5

We illustrate the use of (3.16) for the data of Example 3.4 by reducing the frequency of error ranks into a dichotomy. Define normal errors as a frequency of 10 or less and excessive errors as 11 or more. Then the data are as shown in Table 3.1.

Table 3.1

Assessment	Error frequency		
	Normal	Excessive	Totals
Efficient	10	2	12
Overactive	2	3	5
Totals	12	5	17

We then have

$$t_b = \frac{30 - 4}{\sqrt{[(12)(5)(12)(5)]}}$$

$$= +0.43$$

which is in good agreement with the value of 0.41 found in Example 3.4.

Application to ordered contingency tables

3.15   A similar idea to the one we have used for the 2×2 table can be used to provide a measure of association in a larger contingency table grouped in rows and columns, provided that there are natural orders in these arrays.

Example 3.6

Consider, for example, the data from Stuart (1953) in Table 3.2, showing the grade of vision in the right and left eye of 3242 men aged 30–39. We are interested in the relationship between the grades in right and left eyes and we note that there is an order of rows and columns.

Now goodness of vision (with respect to distance) may be regarded for this purpose as a quality by which the men can be ranked, and any man will have a rank in the two rankings according to right and left eye. We may regard the classification into grades as a rather extensive tying of the 3242 ranks. Thus, for instance, with the right eye, the first 1053 are tied, then the next 782, then the next 893 and finally the last 514. A value of tau calculated for such a table, with due allowances for ties, will measure the association between the vision of right and left eyes for the group of men as a whole.

Table 3.2   Unaided distance vision

Right eye	Left eye				
	Highest grade	Second grade	Third grade	Lowest grade	Totals
Highest grade	821	112	85	35	1053
Second grade	116	494	145	27	782
Third grade	72	151	583	87	893
Lowest grade	43	34	106	331	514
Totals	1052	791	919	480	3242

3.16   To calculate such a coefficient we require the sum S. There will be a positive contribution from each member of any cell with each member lying below it to the right in the table; e..g. from the 821 men in the top left-hand corner there arises a contribution

$$821(494 + 145 + 27 + 151 + 583 + 87 + 34 + 106 + 331).$$

Similarly there will be a negative contribution arising from cells below and to the left; e.g. the score associated with the third-grade left eye and second-grade right eye will be

$$145(-72 - 152 - 43 - 34 + 87 + 331).$$

No score will arise from the bottom line of the table. We find

$$S = 821(1958) + 112(1048) + 85(-465) + 35(-1744)$$
$$+ 116(1292) + 494(992) + 145(118) + 27(-989)$$
$$+ 72(471) + 151(394) + 583(254) + 87(-183)$$
$$= 2\,480\,223.$$

For the denominator of t_b we have the two contributions

$$\tfrac{1}{2}(3242)(3241) - \tfrac{1}{2}(1053)(1052) - \tfrac{1}{2}(782)(781) - \tfrac{1}{2}(893)(892) - \tfrac{1}{2}(514)(513)$$
$$= 3\,864\,293$$

$$\tfrac{1}{2}(3242)(3241) - \tfrac{1}{2}(1052)(1051) - \tfrac{1}{2}(791)(790) - \tfrac{1}{2}(919)(918) - \tfrac{1}{2}(480)(479)$$
$$= 3\,851\,609.$$

Hence

$$t_b = \frac{2\,480\,223}{\sqrt{[(3\,864\,293)(3\,851\,609)]}}$$
$$= 0.643,$$

which provides a measure of the relationship.

3.17 The alternative form t_a has one property (shared by certain other contingency coefficients) of a rather undesirable kind, namely that it cannot attain unity for contingency tables with unequal numbers of rows and columns. This is not a very serious drawback in cases where ties are relatively infrequent, but for heavily tied rankings it may be preferable to use a coefficient for which the limits are attainable.

Consider the maximum score which can be produced by n members arrayed in a table of u rows and v columns. This will be attained when all the observations lie in a longest diagonal of the table and the frequencies in the diagonal cells are as nearly equal as possible. A longest diagonal contains, say, m cells where m is the smaller of u and v. If n is a multiple of m the maximum score is

$$2\left(\frac{n}{m}\right)^2 \{1 + 2 + \cdots + (m - 1)\} = n^2 \frac{m - 1}{m}. \qquad (3.17)$$

When n is not a multiple of m this score is not attainable, but is very nearly so for large n and small m. We therefore define another tau coefficient as

$$t_c = \frac{2S}{n^2[(m - 1)/m]}. \qquad (3.18)$$

We have at once

$$t_c = \frac{n-1}{n} \frac{m}{m-1} t_a. \tag{3.19}$$

In Example 3.6 we considered $n = 3242$ and $m = 4$ so that

$$t_c = \frac{2(2\,480\,223)}{\frac{3}{4}(3242)^2} = 0.629,$$

while the value of t_b is 0.643.

3.18   For a table with the same number of rows as columns (as, for example, in the 2×2 table) t_b can attain unity if the frequencies in the table lie only in the main diagonal. In fact, let the frequencies be $f_1, f_2, \ldots, f_m$. The score S is

$$f_1(n - f_1) + f_2(n - f_1 - f_2) + \cdots + f_m(n - f_1 - f_2 - \cdots - f_m)$$
$$= n \sum f_i - \sum f_i^2 - \sum_{i<j} f_i f_j$$
$$= \tfrac{1}{2}(n^2 - \sum f_i^2).$$

The denominator of t_b for both row and column totals is

$$\tfrac{1}{2}n(n-1) - \sum \tfrac{1}{2}f_i(f_i - 1) = \tfrac{1}{2}(n^2 - \sum f_i^2)$$

and hence t_b is unity. The coefficient t_c is mainly of use when the number of rows and columns is not the same.

3.19   An alternative measure of the relationship for ordered contingency tables that can attain unity for tables with unequal numbers of rows and columns is the Goodman–Kruskal coefficient γ. The numerator is the same S as in t, t_b and t_c, but the denominator is the sum of the number of concordant and the number of discordant pairs. Since the denominator is the actual number of untied pairs, it is obvious that γ can attain unity in the case of perfect agreement. In Example 3.6 the number of concordant pairs is $2\,623\,412$ and the number of discordant pairs is $143\,189$ so that

$$\gamma = \frac{2\,480\,223}{2\,623\,412 + 143\,189} = 0.897$$

as against the values $t_b = 0.643$ and $t_c = 0.629$.

Using the notations of Chapter 1, where P and Q denote the number of concordant and discordant pairs respectively, γ is defined as

$$\gamma = \frac{P - Q}{P + Q}. \tag{3.20}$$

3.20   We conclude this section with another numerical example.

Example 3.7

Dominguez and Vanmarcke (1987) reported the results of a large study to assess the relationship between the competitive structure of Venezuela's markets and basic conditions of supply and demand. Markets were separated into consumer goods and intermediate goods because of the vast differences in conditions of supply and demand. For consumer goods, competitive market structure was assessed according to the following four-point ordinal scale:

1. One or two firms have a dominant position in Venezuela.
2. A small number of firms are divided into distinct competitive customer groups with little mobility among groups.
3. A small number of firms are divided into overlapping competitive customer groups with considerable opportunity for mobility among groups.
4. Many firms are highly competitive with one another for the same customer group.

Also for consumer goods, product life cycle was assessed according to the following four-point ordinal scale:

1. Product newly introduced in Venezuela.
2. Product in growth stage with demand growing faster than Venezuela's economy.
3. Product in mature stage with demand growing slower than Venezuela's economy.
4. Product in declining stage because of substitute products being introduced in Venezuela.

Suppose that the 45 consumer goods companies in the study were classified as follows for competitive market structure and product life cycle.

		Y = Competitive structure				
		1	2	3	4	Total
X = Product life cycle	1	1	0	5	7	13
	2	1	2	6	4	13
	3	3	3	3	1	10
	4	4	5	0	0	9
Total		9	10	14	12	45

Measure the association between competitive structure and product life cycle.

We list each cell in the table and calculate the contributions to P and Q as shown in Table 3.3.

Table 3.3 Calculations for Example 3.7

Cell	Contribution to P		Contribution to Q	
(1, 1)	$1 (2+6+4+3+3+1+5+0+0) = 24$			
(1, 2)	$0 (6+4+3+1+0+0)$	$= 0$	$0 (1+3+4)$	$= 0$
(1, 3)	$5 (4+1+0)$	$= 25$	$6 (3+3+4+5)$	$= 90$
(1, 4)			$7 (1+2+6+3+3+3+4+5+0) = 189$	
(2, 1)	$1 (3+3+1+5+0+0)$	$= 12$		
(2, 2)	$2 (3+1+0+0)$	$= 8$	$2 (3+4)$	$= 14$
(2, 3)	$6 (1+0)$	$= 6$	$6 (3+3+4+5)$	$= 90$
(2, 4)			$4 (3+3+3+4+5+0)$	$= 72$
(3, 1)	$3 (5+0+0)$	$= 15$		
(3, 2)	$3 (0+0)$	$= 0$	$3 (4)$	$= 12$
(3, 3)	$3 (0)$	$= 0$	$3 (4+5)$	$= 27$
(3, 4)			$1 (4+5+0+0)$	$= 9$
(4, 1)				
(4, 2)				
(4, 3)				
(4, 4)				
		$P = 90$		$Q = 503$

We now calculate the corrections for ties from (3.1) and (3.2) as

$$U = [13(12) + 13(12) + 10(9) + 9(8)]/2 = 237$$

$$V = [9(8) + 10(9) + 14(13) + 12(11)]/2 = 238.$$

Hence

$$t_a = \frac{2(90 - 503)}{45(44)} = -0.4172$$

$$t_b = \frac{90 - 503}{\sqrt{45(44)/2 - 237}\,\sqrt{45(44)/2 - 238}} = -0.5488$$

$$t_c = \frac{2(90 - 503)}{45^2(3/4)} = -0.5439$$

$$\gamma = \frac{90 - 503}{90 + 503} = -0.8314$$

Notes and references

The original references for methods of handling tied ranks in rho and tau are 'Student' (1921), Dubois (1939), Woodbury (1940), Kendall (1945, 1947), and Sillitto (1947). The original reference for the case where one ranking is a dichotomy in Section 3.12 is Whitfield (1947). The original references for the Goodman–Kruskal coefficient in Section 3.14 are Goodman and Kruskal (1954, 1959, 1963, 1972). These references are collected in the book by Goodman and Kruskal (1979). A related coefficient proposed by Agresti (1980) is the ratio of the proportion of concordant pairs to the proportion of discordant pairs.

The original references for the tests for two independent samples referred to in Section 3.12 are Wilcoxon (1945) and Mann and Whitney (1947). These are tests for homogeneity between the populations from which the samples are drawn. They are very popular and their properties have been thoroughly investigated in the literature. Terpstra (1952), Kruskal and Wallis (1952), and Jonckheere (1954a) have generalised these tests in certain ways. See Kendall and Stuart (1979, Chapter 31) for additional references.

Another point-biserial rank correlation coefficient can be defined as a function of r_s or $\sum d^2$. Buck (1980) discussed the effect of ties on both of these coefficients.

Stuart (1963) discussed the calculation of rho for contingency tables with ordered classifications. Somers (1962) proposed a measure of association in this situation. Other good references for applications to contingency tables are Stuart (1953), Chapter 33 of Kendall and Stuart (1979), and Goodman (1984). Simon (1978) compared the efficacies of various measures of association for contingency tables.

Problems

3.1 Shipley (1984) reports findings that compare the effectiveness of various criteria used by manufacturing companies in the United States and the United Kingdom to motivate independent intermediaries (distributors, agents, jobbers, manufacturers' representatives and the like). His research hypothesis is that there is a direct association between US and UK companies in their use of these criteria for motivation. The author carried out interview surveys among executives in charge of marketing operations and obtained usable responses from 70 companies in the United States and 59 in the United Kingdom. The data in Table 3.4 represent the percentage of companies that use each criterion for motivation.

Table 3.4

Criterion	Percent of UK firms	Percent of US firms
Financial incentives	69	69
Credit terms	7	20
Advertising/promotions	39	36
Training of employees	51	57
Franchises, dealerships, etc.	8	19
Effective communications	66	64
Inexpensive loans	3	1
Threats to sever relationship	19	26
Other	3	9

(a) Calculate tau both with and without the correction for ties.
(b) Calculate rho both with and without the correction for ties.

Answers: (a) $t_a = 0.9167$; $t_b = 0.9297$ (b) $r_{sa} = 0.9792$; $r_{sb} = 0.9791$.

3.2 The interviews by Shipley (1984) also obtained information about whether or not the criteria used by US and UK companies to select independent distributors are related. His data on the percentage of firms using each criterion are shown below.

(a)   Calculate tau both with and without the correction for ties.
(b)   Calculate rho both with and without the correction for ties.

Table 3.5

Criterion	Percent of UK firms	Percent of US firms
Knowledge of market	83	79
Market coverage	75	79
Sales personnel	49	64
Frequency of sales calls	36	17
Knowledge of product	47	30
Service and stocking	20	23
Service staff	11	27
Enthusiasm	61	50
Previous success	25	67
Costs	25	23
Careers of executives	22	16
Dealings with other firms	10	11
Other	5	9

Answers: (a) $t_a = 0.6410$; $t_b = 0.6536$ (b) $r_{sa} = 0.8214$; $r_{sb} = 0.8207$.

3.3 Hafer and Hoth (1981) investigated the congruity between the attributes sought by employers and the perception of students regarding employers when students are candidates for jobs. The questionnaire listed 26 different attributes to be rated on a five-point Likert scale (1 = no importance, 5 = very important). The subjects were 37 national US business firms and 250 marketing students. The mean ratings of responses to each attribute were calculated for employers and for students. These means were used to obtain the ranks (note the use of average rank for ties) shown in Table 3.6 (1 = least important). Calculate rho and tau both with and without the correction for ties to measure congruence between employers and students.

3.4 Rosenberg, Gibson and Epley (1981) report a study to compare the views of real estate managers and salespersons with regard to job factors that influence the retention of salespeople. Fifty-one brokers/sales managers and fifty-two salespeople from seven US states were surveyed via in-depth interviews to increase the reliability of the results. The data from the paper given in Table 3.7 are the percentages of persons interviewed who considered the job factor a significant one for retention. The respondents could designate more than one factor as most important and therefore the percentage sums can exceed 100%. Calculate both rho and tau, both with and without the correction for ties.

Table 3.6 Ranking of attributes by employers and students

Attribute	Employer ranking	Student ranking
Oral communication	1	1
Motivation	2	2
Initiative	3	9
Assertiveness	4	9
Loyalty	5	9
Leadership	6	12
Maturity	7	7
Enthusiasm	8	3
Punctuality	9	11
Appearance	10	4
Written communication	11	6
Work experience	12	5
Grades	13	16
Disposition	14	13
Extroversion	15	18
Mannerisms	16	15
Willing to relocate	17	14
School reputation	18	20
Social activities	19	22
Knowledge of company	20	17
Community involvement	21	21
Age	22	19
Sports participation	23	26
Hobbies	24	23
Fraternal organization	25	25
Marital status	26	24

Table 3.7

Job factor	Management %	Salespeople %
Money/compensation	24	21
Effective support services	22	15
Good training	18	12
Fair/honest treatment	16	25
Attention to salesperson's needs/ personal relationship	16	15
Recognition	14	10

3.5 Another aspect of the study by Dominguez and Vanmarcke (1987) described in Example 3.7 was to measure the relationship between X = competitive structure and Y = concentration of accounts typically served by large firms for intermediate goods. Competitive market structure was assessed on a two-point scale with the second group combining the last three on the four-point scale outlined in Example 3.7. Concentration of accounts was assessed by the following five-point scale:

1.   More than 6000 accounts
2.   3001–6000 accounts

3.   1001–3000 accounts
4.   101–1000 accounts
5.   Up to 100 accounts.

The 24 firms for intermediate goods were classified as follows for X and Y. Measure the association between X and Y.

		Y = Concentration					
		1	2	3	4	5	Total
X = Competitive structure	1	8	3	2	0	0	13
	2	2	3	1	1	4	11
Total		10	6	3	1	4	24

Answer: $t_b = 0.488$.

3.6   The study by Davis (1970) discussed in Example 1.5 and Problem 1.3 was also interested in learning whether husbands and wives agree about the influence pattern of wife dominance. The data given were the percentage of husbands who answered (4) wife has more influence than husband or (5) wife decided, on the 12 questions asked about purchasing both furniture and automobile, and the percentage of wives who answered (4) or (5) as above. Measure the agreement between husband and wife concerning wife dominance using the correction for ties.

Table 3.8

Decision	Wife	Husband	Decision	Wife	Husband
1	2	3	7	20	31
2	2	3	8	30	39
3	4	1	9	33	40
4	0	8	10	44	64
5	1	9	11	54	72
6	12	25	12	74	82

Answer: $r_{sb} = 0.874$.

3.7   Two additional research studies were presented in Davis (1970). One study was concerned with the influence pattern of husband dominance measured by the percentage of each sex who answered (1) husband decided or (2) husband has more influence than wife, on the 12 questions. The second study was concerned with the pattern of joint dominance as measured by the percentage of each sex who answered (3) husband and wife have equal influence, on the 12 questions. For each set of data (Table 3.9), compute the Spearman rank correlation with correction for ties.

Table 3.9

Question	(a) Husband dominance		(b) Joint dominance	
	Wife	Husband	Wife	Husband
1	68	68	30	29
2	59	62	39	35
3	62	62	34	37
4	50	60	50	32
5	47	41	52	50
6	25	25	63	50
7	17	22	63	47
8	18	16	52	45
9	6	7	61	53
10	4	3	52	33
11	2	2	45	26
12	2	2	24	16

Answers: (a) $r_{sb} = 0.991$ (b) $r_{sb} = 0.786$.

3.8 Jackman (1985) asked a group of subjects to rate 308 words with respect to imagery and part of speech. Imagery ratings were given on a 7-point scale where 1 = very abstract, 7 = very concrete. Part-of-speech ratings were given on a 5-point scale where 1 = always a noun, 2 = usually a noun, 3 = both a noun and a verb, 4 = usually a verb, 5 = always a verb. The frequency distribution of the average ratings given to the 308 words by all subjects combined are shown in Table 3.10. Calculate Kendall's tau as a measure of agreement between imagery rating and part-of-speech rating using the correction for ties.

Table 3.10

Part-of-speech rating	Imagery rating					
	1-1.99	2-2.99	3-3.99	4-4.99	5-5.99	6-7.00
1-1.99	4	16	10	32	47	32
2-2.99	1	7	12	15	31	11
3-3.99	3	13	20	18	10	0
4-5.00	1	11	8	5	1	0

Chapter 4

Tests of significance

4.1 The rankings with which we deal in practice are usually based on a set of individuals which themselves are only samples from a much larger population. It is of some interest to be able to measure the relationship between mathematical and musical ability in a given class of children; but it is of much greater interest to be able to say, if this class is chosen at random from a certain population of children, how far the results for the sample throw light on the relationship in that population. In this chapter we shall consider the question: given a value of a rank correlation in a sample, can we conclude that there exists correlation in the population from which the sample was chosen? In short, we shall try to *test the significance* of observed rank correlations in the special sense of the statistical theory of sampling.

4.2 Suppose that in the present population there is no relationship between the two qualities under consideration. Then if a sample is chosen at random, any order for the quality X is just as likely to appear with a given order for Y as any other X-order. If we choose some arbitrary order for Y (it does not matter which, so we will take the natural order $1, \ldots, n$), then all the $n!$ possible rankings of the numbers 1 to n for X are equally probable. Each accordingly has the probability $1/n!$ (For the present we confine ourselves to the untied case.)

   Now to each of the possible arrangements of the X-ranking there will correspond a value of t or r_s. The totality of such values, $n!$ in number, may be classified according to the actual value of t or r_s, ranging from -1 to $+1$, in what is called a *frequency distribution*. This distribution is fundamental to our present investigation and we consider it in some detail.

4.3 For a ranking of four there are 24 possible arrangements. If we write down all arrangements and calculate S for each one paired with the natural order 1, 2, 3, 4, we find the frequency distribution shown in Table 4.1. Thus the greatest frequency of the values of S is 6, attained when $S = 0$. The distribution is symmetrical about this value, and as S becomes greater in absolute value the frequencies decrease to unity.

4.4 The corresponding distribution for $n = 8$ is shown in Table 4.2 (we show only zero or positive values of S, the negative values being given by symmetry).

Table 4.1

Value of S	Frequency of rankings with the assigned value of S
-6	1
-4	3
-2	5
0	6
2	5
4	3
6	1
Total	24

Table 4.2

Values of S	Frequency of rankings with the assigned value of S
0	3 826
2	3 736
4	3 450
6	3 017
8	2 493
10	1 940
12	1 415
14	961
16	602
18	343
20	174
22	76
24	27
26	7
28	1
Total (of whole distribution):	40 320

Once again we find a maximum value at $S = 0$ and a steady decrease in the frequency as S increases. Figure 4.1 shows a smooth curve that approximates a frequency polygon of this distribution.

4.5 In the next chapter we shall see how to derive these distributions for various values of n. For the present we state without proof:

(a) The distributions are always symmetrical. If $\frac{1}{2}n(n-1)$ is even, S can take only even values and there is a maximum frequency for $S = 0$. If $\frac{1}{2}n(n-1)$ is odd, S can take only odd values and there is a pair of maximum frequencies at $S = \pm1$.

(b) The frequencies decrease steadily from the maximum to the value of unity for $S = \pm\frac{1}{2}n(n-1)$.

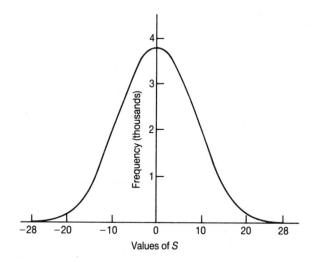

Fig. 4.1.

(c)   As *n* increases, the shape of the frequency polygon tends to that of the *normal curve*

$$f(x) = \frac{1}{\sigma\sqrt{(2\pi)}}\,e^{-x^2/2\sigma^2}. \tag{4.1}$$

For *n* greater than 10 this curve provides a satisfactory approximation to the polygon with mean zero and standard deviation σ where

$$\sigma^2 = \tfrac{1}{18}n(n-1)(2n+5). \tag{4.2}$$

If the normal curve with this standard deviation for $n = 8$ were drawn in Fig. 4.1 it would fall so close to the polygon as to be barely distinguishable on this scale.

4.6   The tendency of the distribution of frequencies to normality is a very useful property which enables us to avoid the calculation of the actual distribution for $n > 30$. For $n \leq 30$ the exact distribution is given in Appendix Table 1. For $n \leq 10$, this table shows the proportionate frequencies (i.e. the actual frequencies divided by $n!$) for which S attains or exceeds a specified value. For $11 \leq n \leq 30$, the values of t for selected quantiles are given.

The significance of *t*

4.7   Let us now consider the use of these distributions in testing the null hypothesis of independence or $H\colon \tau = 0$, where τ is the parameter that measures the coefficient of concordance in the population from which our random sample is drawn. The test is based on the value of t or equivalently on the corresponding value of S, since one is the multiple of the other.

If there is no connection between the two qualities, a pair of rankings chosen at random will give some value of S lying between the limits $\pm\frac{1}{2}n(n-1)$. The greater part of such values will cluster round the value zero. We shall adopt the following criterion: if the observed value of S is such that it is very improbable that such a value *or greater* in absolute value could have arisen by chance, we shall reject the null hypothesis that the two qualities are independent. This amounts to saying that if an observed value of S lies in the 'tails' of the distribution of S away from its mean value we shall reject the hypothesis. Just where we draw the line to decide on what are the tails or what is very improbable is a matter of convention. Many people regard a probability of 0.05 or one of 0.01 as small and one of 0.001 as very small. We sometimes speak of a '5 per cent probability level' of S, meaning that value of S which is attained or exceeded with probability 0.05, and similarly for a '1 per cent probability level' or a '0.1 per cent probability level'. The corresponding values of S may be termed, for example, the '5 per cent significance point'. To say that an observed S lies beyond the 5 per cent significance point means that the probability of arriving at such a value or greater (in absolute value) is less than 0.05.

If we suspect beforehand that there is positive correlation we may prefer to consider the probability that S lies in the upper tail only; and similarly, if negative correlation is expected we may confine attention to the lower tail. This would amount to calculating probabilities that S attained or exceeded some given value or fell short of some value, as the case might be, instead of attaining or exceeding some figure regardless of sign. In the former case the appropriate alternative hypothesis is $A_+: \tau > 0$, and in the latter case the alternative is $A_-: \tau < 0$. In each case, we have what is called a one-sided alternative or a directional alternative.

Example 4.1

In a random sample giving a ranking of 10 persons regarding ability in mathematics and music, we find a value of $t = -0.11$ and $S = -5$. Is such a value significant?

From Appendix Table 1 we see that the proportion of rankings which give a value of -5 or less and $+5$ or more is twice 0.364, namely 0.728. This is quite large and we need not reject the null hypothesis of independence of the two qualities. In other words, the observed value is not significant. It could well have arisen by chance from a population in which mathematical and musical ability are unrelated.

Had the observed value of t been $+0.56$, corresponding to a value of $S = 25$, we find the probability that $|S| \geq 25$ from Appendix Table 1 is $2(0.014) = 0.028$. This is very small, and we should have concluded that the abilities were not independent in the parent population.

Now suppose that we are interested in the one-sided alternative $A_+: \tau > 0$ in this same example where $S = 25$ and $t = 0.56$ for $n = 10$. Since a large positive value of S or t makes us believe this directional alternative is true for the parent population, we should report the proportion of rankings which give an S greater than or equal to 25 in that direction only. Then we have

$$\text{Prob}(S \geq 25) = 0.014,$$

which of course is one-half of the result reported above for the two-sided alternative where we asked only if the value of t or S is significant.

When we observed $S = -5$ and $t = -0.11$ for $n = 10$, the appropriate directional alternative is A: $\tau < 0$ because S and t are negative. Thus here we report

$$\text{Prob}(S \leq -5) = 0.364.$$

P-values

4.8   The probabilities reported in the tests of significance in Example 4.1 are usually called *P*-values. Suppose in the general case that we observe S_0 for the value of S or t_0 for the value of t. The *P*-value to report for the two-sided or non-directional alternative $A: \tau \neq 0$ is

$$\text{Prob}\{|S| \geq |S_0|\},$$

which is twice the value of $\text{Prob}\{S \geq |S_0|\}$ given in Appendix Table 1 for S_0. The *P*-value to report for the one-sided alternative $A_+: \tau > 0$ is

$$\text{Prob}\{S \geq S_0\}$$

for S_0 positive, and the appropriate *P*-value for $A_-: \tau < 0$ is

$$\text{Prob}\{S \leq S_0\}$$

for S_0 negative.

If we have found the appropriate *P*-value for a given test of significance and want to make a decision at significance level α, the decision should always be

Reject H if *P*-value $\leq \alpha$, and
Do not reject H if *P*-value $> \alpha$.

When n is less than or equal to 10, Appendix Table 1 gives the exact *P*-values. When n is greater than 10 but not greater than 30 we can use Appendix Table 1 to find significance points or ranges of *P*-values.

When n is greater than 10 we could also use the table of areas under the standard normal curve given in Appendix Table 3 to find an approximation to the exact *P*-value of S or t in a test of significance. Each area given in Appendix Table 3 is the probability that a given positive multiple of the standard deviation will be attained or exceeded, i.e. a right-tail *P*-value, for an observed positive standard normal variable z. The standard normal test statistic for $H: \tau = 0$ based on S or t is

$$z = \frac{3S\sqrt{2}}{\sqrt{n(n-1)(2n+5)}} = \frac{3t\sqrt{n(n-1)}}{\sqrt{2(2n+5)}} \qquad (4.3)$$

because the standard deviation of t is $2/n(n - 1)$ times that of S given in (4.2). The appropriate P-values for one-sided and two-sided alternatives are the same for z as for t but taken from Appendix Table 3.

In the first part of Example 4.1 where $S = -5$, $t = -0.11$, and $n = 10$, we use (4.3) to calculate $z = -0.45$. Appendix Table 3 shows that the proportion of rankings which give a value of z which is -0.45 or less is 0.3264 so that the approximate P-value is $2(0.3264) = 0.6528$, to be compared with the exact value 0.728 which we found previously. This approximation is not very accurate for $n = 10$ and is not generally recommended for $n < 30$. It can however be improved by a continuity correction.

Continuity correction for *S*

The normal distribution is continuous, while the exact distributions of both t and S are discrete or discontinuous. The approximation based on the normal distribution can be improved by incorporating a continuity correction in the test statistic z. This is not easily accomplished for z as a direct function of t as in the second expression of (4.3) because the differences in possible values for t are not constant. However, the possible values of S are always only even integers or only odd integers depending on whether $n(n - 1)/2$ is even or odd. Since the possible values always differ by two units, we shall regard each S value as spread uniformly between $S - 1$ and $S + 1$, instead of concentrated wholly at S. In other words, we shall subtract unity from the observed S if it is positive and add unity if it is negative in calculating the numerator of z in the first expression in (4.3).

In the first part of Example 4.1 where $S = -5$ and $n = 10$, we shall see how the continuity correction improves the normal approximation. Since S is negative, we add one to obtain -4 to substitute for S in the first expression in (4.3). The result is $z = -0.36$ with an approximate P-value of 0.7188, which is much closer to the exact value of 0.728. We emphasise the importance of the continuity correction by another example.

Example 4.2

In a pair of rankings of $n = 20$ the value of S observed is 58 and accordingly $t = 0.31$. Is this significant? Appendix Table 1 shows that the one-tailed P-value is between 0.025 and 0.05, and the corresponding two-tailed P-value is then between 0.05 and 0.10. Our table does not enable us to be more precise.

Let us use the normal approximation with a continuity correction for this problem. Since S is positive, we subtract one to get 57 and (4.3) yields $z = 1.85$. From Appendix Table 3 the probability of a normal deviate of less than 1.85 is 0.9678 so that the two-tailed P-value is about $2(1 - 0.9678) = 0.064$. This is small, but not very small, so we simply report the P-value and make no definitive conclusion.

As another example, if $n = 9$ and $S = 20$, we have

$$Z = \frac{S - 1}{\sigma} = \frac{19}{9.592} = 1.981.$$

The probability that this will be attained or exceeded in absolute value is seen, from Appendix Table 3, to be about 0.048. The exact value, from Appendix Table 1, is 0.044. Had we made no correction for continuity we should have found a *P*-value of 0.037.

Ties

4.9   When ties are present the above formula (4.2) for the standard error of S requires some modification. If there are ties of extent u in one ranking and v in the other, then the variance of the distribution obtained by correlating one ranking with all $n!$ possible arrangements of the other is given by

$$\text{var } S = \tfrac{1}{18}\{n(n - 1)(2n + 5) - \sum u(u - 1)(2u + 5) - \sum v(v - 1)(2v + 5)\}$$

$$+ \frac{1}{9n(n - 1)(n - 2)} [\sum u(u - 1)(u - 2)][\sum v(v - 1)(v - 2)]$$

$$+ \frac{1}{2n(n - 1)} [\sum u(u - 1)][\sum v(v - 1)]. \tag{4.4}$$

If only one ranking contains ties, so that all vs are zero,

$$\text{var } S = \tfrac{1}{18}[n(n - 1)(2n + 5) - \sum u(u - 1)(2u + 5)]. \tag{4.5}$$

We shall prove these results in the next chapter.

4.10   As for the untied case, the distribution of t for any fixed number of ties tends to normality as n increases, and there is probably little important error involved in using the normal approximation for $n \geq 10$, unless the ties are very extensive or very numerous, in which case a special investigation may be necessary. References are given at the end of this chapter.

Example 4.3

Consider the two rankings of 12:

X	$1\frac{1}{2}$	$1\frac{1}{2}$	3	4	6	6	6	8	$9\frac{1}{2}$	$9\frac{1}{2}$	11	12
Y	$2\frac{1}{2}$	$2\frac{1}{2}$	7	$4\frac{1}{2}$	1	$4\frac{1}{2}$	6	$11\frac{1}{2}$	$11\frac{1}{2}$	$8\frac{1}{2}$	$8\frac{1}{2}$	10

We find

$$S = 8 + 8 + 1 + 5 + 5 + 5 + 5 - 3 - 2 + 1 + 1 = 34.$$

In the first ranking there are ties of extent 2, 2, 3 and in the second of extent

2, 2, 2, 2. From (4.4) we then have

$$\text{var } S = \tfrac{1}{18}[(12)(11)(29) - 6(2)(1)(9) - (3)(2)(11)] + 0$$

$$+ \frac{1}{2(12)(11)} [2(2)(1) + (3)(2)]\{4(2)(1)]$$

$$= 203.30.$$

Thus with a correction for continuity

$$Z = \frac{33}{\sqrt{203.30}} = 2.31.$$

The chance of attaining or exceeding such a value in absolute magnitude is about 0.021. This is small, and we incline to attribute significance to the value of S.

4.11 If one ranking degenerates to a dichotomy, as in Section 3.13, with x and $n - x = y$ members and ties in the other ranking typified by u, we find, on substitution in (4.4), the equation

$$\text{var } S = \frac{xy}{3n(n - 1)} [n^3 - n - \sum (u^3 - u)]. \tag{4.6}$$

Finally, if both rankings become dichotomies, as in 3.14, we find on substitution in (4.6)

$$\text{var } S = \frac{xypq}{n - 1}. \tag{4.7}$$

4.12 These equations provide us with tests of significance of the appropriate t_a, t_b or t_c coefficients, or rather of the values of S from which they are derived. There is, however, one difficulty in regard to corrections for continuity:

(a) In the case of a dichotomy and an *untied* ranking the interval between successive values of S is two. The appropriate deduction from S for continuity is one-half of the value, namely unity.

(b) For a dichotomy and a ranking composed entirely of ties of the same extent u the interval is $2u$ and the appropriate deduction for continuity is u.

(c) When both variates are dichotomies the interval is n and the deduction for continuity is $\tfrac{1}{2}n$.

(d) If one variate is dichotomised and the other contains ties of varying extents, there will be varying differences of interval between successive values of S in various parts of the range. In such a case we may use an approximate method, as in Example 4.5 below.

Example 4.4

In Example 3.5 we found a value

$$t_b = 0.43$$

based on $S = 26$. For the variance we have, from (4.7),

$$\sigma_s^2 = \text{var } S = \frac{12(5)(12)(5)}{16}$$

$$\sigma = 15.$$

The correction for continuity is $17/2$ and thus we have

$$Z = \frac{S - 8.5}{15} = 1.167.$$

The probability that this is attained or exceeded in absolute value is 0.24, and the value of t_b is not significant.

Example 4.5

For the same data but with one variate not dichotomised we have found, in Example 3.4, $t_b = 0.41$, based on $S = 34$. For the variance we have, from (4.6),

$$\sigma^2 = \text{var } S = \frac{12(5)}{3(17)(16)} [17^3 - 17 - (4^3 - 4) - (2^3 - 2)$$

$$- (5^3 - 5) - (3^3 - 3)]$$

$$= 344.6$$

$$\sigma = 18.56.$$

Now for the continuity correction we note that the X-ranking 'jumps' from $2\frac{1}{2}$ to 5 and this involves an interval of 5 in the S score; for if we replace one of the E's corresponding to an X-rank by the O corresponding to the X-rank of 5, S is reduced by 5, the first five scores being $4 + 4 + 4 - 9 + 4$ instead of $5 + 5 + 5 + 5 - 8$. Similarly the 'jump' from 5 to $6\frac{1}{2}$ gives an interval of 3 and so on. We can estimate the mean interval without calculating each individual interval in this way. The total of the S-score intervals is twice the number of members in the ranking less the extent of the ties involving the first and last member. In our present case this is $34 - 4 - 1 = 29$, and the mean interval is thus $29/6 = 4.833$. We take half this as the correction for continuity and thus have

$$Z = \frac{S - 2.416}{18.56} = 1.70,$$

giving a probability that S will be equalled or exceeded absolutely as 0.089. The value of t_b is still not significant.

It is worth noting that in the previous example we found, for the 2×2 table, a value of $t_b = 0.43$, which was not significant. In the present case we find $t_b = 0.41$, a slightly smaller value which has a probability of 0.089 of being exceeded in absolute value, as against 0.24 for the value based on the 2×2 table. This is not a discrepancy. In this example we have not dichotomised the second ranking but have taken all the ranks into account. Our method gives more play to values which might be assumed by chance, and hence the probability of exceeding a given value in this more extended field may well differ from the value obtained in the more restricted domain of the double dichotomy.

The significance of r_s

4.13 Just as for t, the set of $n!$ values obtained by correlating all possible (untied) rankings with an arbitrary ranking will provide a set of r_s values which may be used to test the significance of that coefficient. The distribution of r_s is symmetrical about 0 and tends to normality for large n but approaches normality much slower than t does.

Appendix Table 2 gives the exact probability distribution of $\sum d^2$ for $n \leq 16$ as all left-tail P-values which do not exceed 0.5 for $r_s < 0$ or equivalently $\sum d^2 > n(n^2 - 1)/6$. By symmetry, we can find right-tail P-values for $\sum d^2 < n(n^2 - 1)/6$ and $r_s > 0$ by entering this table with $n(n^2 - 1)/3 - \sum d^2$. For example, with $n = 4$, $\sum d^2 = 16$, we read out a left-tail P-value of 0.208 directly for $r_s = -0.60$. But with $n = 4$, $\sum d^2 = 8$, we enter the table with $20 - 8 = 12$ and read out a right-tail P-value of 0.458 for $r_s = 0.20$.

For $17 \leq n \leq 35$, Appendix Table 2 gives the values of r_s for selected quantiles. For example, if $n = 25$ and we have $r_s = 0.398$, the right-tail P-value is 0.025, which is also the left-tail P-value for $r_s = -0.398$.

4.14 For $n > 35$, it is probably accurate enough to use the normal approximation. The variance of r_s is

$$\text{var}(r_s) = \frac{1}{n - 1}. \tag{4.8}$$

The corresponding expression for $\sum d^2$ is

$$\text{var}(\sum d^2) = \left(\frac{n^3 - n}{6}\right)^2 \frac{1}{n - 1} = n^2(n - 1)(n + 1)^2/36. \tag{4.9}$$

Thus the test statistic for an approximate test of the null hypothesis of independence or $H: \rho_s = 0$, where ρ_s is the parameter representing the rank correlation coefficient in the population from which the sample is drawn, is

$$z = r_s\sqrt{n - 1} = \sqrt{n - 1}\left[1 - \frac{6 \sum d^2}{n^3 - n}\right], \tag{4.10}$$

with corresponding P-values obtained from Appendix Table 3.

For $19 \leq n < 30$, a more accurate approximate test is provided by the statistic

$$t_{n-2} = \frac{r_s \sqrt{n-2}}{\sqrt{1 - r_s^2}} \tag{4.11}$$

and the Student's t-distribution with degrees of freedom $df = n - 2$. Actually, this test can be used with reasonable accuracy for $n > 10$.

Example 4.6

In a ranking of 20 the observed value of $\sum d^2$ is 840. We then have

$$r_s = 1 - \frac{6(840)}{20^3 - 20}$$

$$= 0.3684.$$

Appendix Table 2 indicates that the right-tail P-value is between 0.05 and 0.10. The test statistic from (4.11) is $t_{18} = 1.68$ with 18 degrees of freedom. The probability that this is attained or exceeded in absolute value is between 0.10 and 0.20. The corresponding test statistic from (4.10) is $z = 1.61$. The probability that this is attained or exceeded in absolute value is 0.1074 from Appendix Table 3.

Figure 4.2 shows the distribution of $\sum d^2$ as a frequency polygon for $n = 8$. The saw-like appearance is unusual, but the correspondence with a normal curve is moderately good except in the tails. The fit is not good enough for the purpose of tests of significance because then we are mainly interested in the tails.

Continuity correction for $\sum d^2$

4.15 If we wish to make a continuity correction for the test of significance based on the normal approximation, we note that $\sum d^2$ can take on only even integer values between 0 and $(n^3 - n)/3$, with a mean of $(n^3 - n)/6$. Therefore, to make the correction, we shall substract unity from the observed $\sum d^2$ if it is less than the mean $(n^2 - n)/6$, and add unity if it exceeds $(n^3 - n)/6$ in calculating the numerator of z in the second expression in (4.10).

In Example 4.6, the observed $\sum d^2$ is 840, which is less than the mean 1330 for $n = 20$. Therefore we should substract unity and substitute 839 for $\sum d^2$ in (4.10). The result is $z = 1.61$, exactly as before without the continuity correction.

Note that it is not easy to incorporate a continuity correction in the expressions of (4.10) and (4.11) that involve r_s because the differences in possible values of r_s are not constant. This is a disadvantage of r_s as compared to $\sum d^2$.

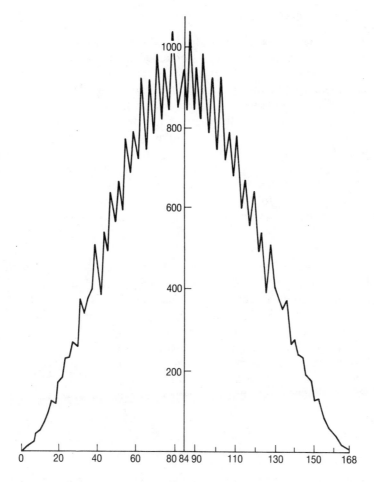

Fig. 4.2.

Corresponding to the corrections for continuity to S given in Section 4.12 we have the following corrections to $\sum d^2$ in tied cases:

(a) For a dichotomy against an untied ranking the interval between successive values is n and we therefore deduct $\frac{1}{2}n$ from $\sum d^2$.

(b) For a dichotomy against a ranking composed entirely of ties of extent u the interval is nu and the appropriate deduction for continuity is $\frac{1}{2}nu$.

(c) For a double dichotomy the deduction for continuity is $\frac{1}{4}n^2$.

It is worth noting, however, that for a double dichotomy r_{sb} is the same as t_b and the formulae for the latter would customarily be used in calculation.

If ties are present no modification is required in the variance of r_s, which remains as $1/(n-1)$.

Tests in the non-null case

4.16   The tests we have used up to this point are based on the distribution of correlations in a population of values obtained by permuting the ranks in all possible ways. This, effectively, is a test of the hypothesis that the qualities under consideration are independent in the parent population. Our test will then show whether an observed correlation is significant of a departure of the parent correlation *from zero*. But we may also wish to test correlations from a rather different viewpoint, or rather, to assign limits to the parent value in some probabilistic sense. For instance, suppose we arrive at a value of *t* equal to 0.6 and it is found to be significant. Can we say between what values the true correlation probably lies? Again, if we have a further value from a different sample equal to 0.8, also significant, can we say that the second is significantly greater than the first, or is the difference such as may have arisen by chance?

4.17   Suppose the whole population of N is ranked according to the first variate in the order $1, \ldots, N$. This can evidently be done without loss of generality. Suppose the population is laid out in a line in this order, the ranks according to the second variate being p_i for the ith member. Then in accordance with our usual technique we may calculate a value of τ for this population.

Now suppose that a sample of n members is chosen from the N. These members will be in the natural order according to the first variate. We may then calculate a value for the sample t. We merely write t instead of τ to denote that we are considering a sample value instead of a population value. To each possible sample there will correspond a value of t, and since there are

$$\binom{N}{n} = \frac{N!}{n!(N-n)!}$$

samples there will be an equivalent number of values of t corresponding to all possible samples.

4.18   In the next chapter we shall show that for any parent whatsoever the distribution of t tends to normality as n, the sample size, increases, provided that the parent τ is not too near to unity and certain non-restrictive conditions are obeyed. We shall also show that the mean value of this distribution is τ. So far, so good, but we then encounter difficulties. If the standard deviation of the distribution were dependent only on τ we could easily test the observed value in the desired manner; but in fact the standard deviation depends on other unknown quantities.

4.19   A simple illustration will emphasise the point. Consider the ranking of 9 (according to the second variate):

5   2   3   1   6   7   8   9   4

There are $\binom{9}{3}$ = 84 possible samples of 3 from this 'population'. If they are written down and P evaluated for each we find the distribution shown in Table 4.3.

Table 4.3

Values of P	Frequency
0	2
1	15
2	34
3	33
	Total: 84

The mean of this distribution will be found to be 13/6 and hence the mean value of the 84 values of t is

$$E(t) = \frac{26/6}{3} - 1 = 0.44.$$

It will be found that, for the parent τ, $P = 26$ and hence

$$\tau = \frac{52}{36} - 1 = 0.44 = E(t),$$

verifying our statement that the mean value of t is the parent τ. The ranking

$$1 \quad 2 \quad 5 \quad 9 \quad 3 \quad 6 \quad 7 \quad 8 \quad 4$$

also has $\tau = 0.44$, but the distribution of P in samples of 3 is now as shown in Table 4.4

Table 4.4

Values of P	Frequency
0	3
1	16
2	29
3	36
	Total: 84

Again the mean value of t is equal to the parent τ, but the distribution of P in the second case is different from that in the first, and its variance is 0.734 against 0.639 in the first case.

4.20 We are therefore in the difficulty that unless we know something about the arrangement of the ranks in the parent (knowledge which is usually lacking) we cannot express the variance of t in terms of known factors. We shall, however, show in the next chapter that for any parent the variance of t

cannot exceed a certain value, namely

$$\text{var } t \leq \frac{2}{n}(1 - \tau^2). \tag{4.12}$$

This result will give us a test which is on the safe side. This expression also holds good when ties are present. A corresponding approximate expression for the variance of r_s is

$$\text{var } r_s \leq \frac{3}{n}(1 - \rho^2) \tag{4.13}$$

for large n, but this is not known to be valid for the tied case. The argument will be clear from an example.

Example 4.7

In a ranking of 30 a value of t is found to be 0.816. Assuming this ranking to be a random sample, what can be said about the value of τ in the parent?

For samples of 30 the distribution of t may be taken to be normal, and its mean may be estimated as 0.816. From (4.12) we find that

$$\text{var } t \leq \tfrac{2}{30}[1 - (0.816)^2] = 0.022\,276,$$

giving a standard error of 0.149 *or less*.

Now the probability of a deviation from the mean of 1.96 times the standard deviation or greater (in absolute value) is 0.05. We may therefore say that at the worst, the probability is 0.95 that the true value of τ lies within $(0.149)(1.96) = 0.292$ of 0.816, i.e. that the probability is less than or equal to 0.05 that the true value lies outside the range 0.524 to 1.0. Instead of setting limits to the value in the usual statistical manner with an assigned degree of probability, we set limits with a maximum to the degree of probability; or, to put it another way, we set *outside* limits to the range of the parent τ. This may make our inference unnecessarily stringent, but we err on the safe side in the sense that we run no danger of attributing significance to non-significant results, though we may in some cases fail to discern significance where it really exists.

4.21   We have stated the argument above by relation to standard errors. In order to apply (4.12), however, we need to replace the unknown parent value τ by the sample value t, as has been done in the previous example. Such a procedure may be avoided by recourse to the theory of confidence intervals. If z is the normal deviate corresponding to a probability level of P per cent (i.e. if the probability is $0.0P$ that there will occur on random sampling a deviation from the mean of z times the standard deviation or greater in absolute value), then, to this probability at the most, we may assert that

$$\tau_1 \leq \tau \leq \tau_2 \tag{4.14}$$

where τ_1, τ_2 are the roots of

$$t - \tau = x\sqrt{\left[\frac{2}{n}(1 - \tau^2)\right]},$$

given by

$$\tau = \frac{t \pm x\sqrt{(2/n)}\sqrt{[1 + (2x^2/n) - t^2]}}{1 + (2x^2/n)}. \tag{4.15}$$

Example 4.8

For example, in the data of Example 4.7, $n = 30$, $t = 0.816$. The normal deviate corresponding to a probability of 0.05 is 1.96. Substitution in (4.15) then gives

$$\tau = \frac{t \pm 0.50607\sqrt{(1.2561 - t^2)}}{1.2561}$$

$$= 0.34 \quad \text{or} \quad 0.96.$$

Thus we may assert that τ lies between 0.34 and 0.96, being sure of correctness in at least 95 per cent of the cases on the average. This is a more accurate method than that of Section 4.20. The limits are, of course, still maxima.

Example 4.9

In a sample of 20 a value of $t = 0.8$ is observed. A second sample of 20 gives a value of 0.6. Is there any indication that the samples are from different material, or could the different values of t arise by chance?

For a maximum to the variance in the first case we have

$$\text{var } t = \frac{2}{n}(1 - t^2)$$

$$= 0.036,$$

giving a standard error of 0.19. The second value differs by about this amount, and consequently we cannot attribute significance to the result.

Alternatively, we might argue that the variance of the second value is 0.064. Thus, a maximum to the variance of their difference (being equal to the sum of the variances) is 0.100, the corresponding standard error being 0.32. This is greater than the actual difference of 0.2, and again we conclude that the difference is not significant.

Generally, if we have a number of values of S (even from rankings of different extents) we may add them together and test the significance of the whole set against the sums of the variances of the individual rankings. The validity of this procedure rests on the fact that the variance of the sum of *independent* variables is the sum of their variances. It has been found very useful in field work.

4.22   The foregoing examples illustrate one rather disappointing feature of rank correlation coefficients, namely the comparatively large standard error which they possess. Whatever τ may be, the standard error of t is of the order of $\sqrt{(2/n)}$. This shortcoming, however, is a feature of most correlation coefficients. The standard error of the product-moment coefficient in normal samples, for instance, is $(1 - \rho^2)/\sqrt{n}$ and thus is, in general, of order $1/\sqrt{n}$. It is clearly impossible to locate the parent correlation very closely unless the ranking contains 30 or 40 members. This provides a useful caution against attributing reality to correlation coefficients calculated from rankings of small extent, unless several sample values are available.

For instance, with a ranking of 32 members the maximum standard error is $\frac{1}{4}\sqrt{(1 - \tau^2)}$ and if t is near zero we cannot, on the above basis, locate the parent τ in a narrower range than twice the standard error, or ± 0.5. If t is, say, 0.8 the range becomes narrower, but the band of doubt surrounding the true value still extends from 0.5 to 1.0.

4.23   It is not possible to improve very much on the maximum limits given by (4.12) or (4.14), as we shall show in the next chapter. We shall see, however, that very substantial improvements are possible if the original rankings are available and the investigator has patience to carry out the necessary arithmetic.

Applications to time-series data

4.24   As mentioned in Section 1.24, tau and rho are useful in describing relationships for time-series data. We let the X-variable denote time and the Y-variable represent the measurements taken over time and compute tau or r_s for these observations (X, Y). If the null hypothesis is that the $Y_1, Y_2, ..., Y_n$ are independent and identically distributed, the null distributions of r_s and t are exactly the same as described in this chapter. Suppose the alternative hypothesis is that the Y measurements exhibit a trend. A large positive value of t or r_s indicates that the measurements tend to become larger as time goes on, i.e. the data exhibit a positive trend. A large negative value of t or r_s indicates the opposite, i.e. data exhibiting a negative trend over time.

The test for trend based on tau is known as the Mann test, and the one based on rho is the Daniels test.

Example 4.10

For the 1970 draft of the US armed forces, the selection was to be made at random according to their day (not year) of birth from all men aged at least 19 but less than 26. The 366 days (including February 29) were each assigned a three-digit number (for example, September 14 was number 001) to represent priority by the following process. The January dates were inserted into 31 blank capsules and placed in a drum. The same procedure was followed for

February, then March, and so on, with a mixing each time a month was added to the drum. The order of induction of elegible men was based on the order of the dates that appeared in a random drawing of these capsules. Since September 14 was drawn first, all eligible men born on September 14 were drafted first. The data in Table 4.5 show the ranking of the average lottery numbers assigned by month (1 = smallest average). The randomness of the drawing was criticised on the basis that later months tended to have lower lottery numbers than earlier months and that this happened because the capsules for later months were on top of the drum and not thoroughly mixed in. Investigate this criticism by determining whether the ranking of average lottery numbers by month shows a significant negative trend.

Table 4.5

January	8		July	5
February	9		August	4
March	12		September	3
April	10		October	6
May	11		November	2
June	7		December	1

We number the successive months from 1 to 12 to represent the X ranks and list the corresponding ranks of average lottery numbers to calculate r_s and t as shown in Table 4.6

Table 4.6

Month rank	Lottery rank	d^2	P	Q
1	8	49		
2	9	49	4	7
3	12	81	3	7
4	10	36	0	9
5	11	36	1	7
6	7	1	0	7
7	5	4	0	6
8	4	16	1	4
9	3	36	1	3
10	6	16	1	2
11	2	81	0	2
12	1	121	0	1
		526	11	55

The results are $r_s = -0.8392$ with a left-tail P-value from Appendix Table 2 as 0.000 57 and $t = -0.6667$ with a left tail P-value less than 0.005 from Appendix Table 1. In each case we clearly reject the null hypothesis of no trend and conclude that a significant negative trend exists. See Fienberg (1971) for additional details and comments.

Notes and references

For the null distribution of tau see Kendall (1938), Dantzig (1939), Daniels (1944), Sillitto (1947), Moran (1950b), Silverstone (1950) and David, Kendall and Stuart (1951). More extensive tables were developed by Kaarsemaker and Van Wijngaarden (1952, 1953), Dunstan, Nix and Reynolds (1979). Other papers concerned with the null distribution are Best (1973), Best (1974), Best and Gipps (1974), Nijsse (1988). The distribution of tau in the presence of ties is discussed in Sillitto (1947), Smid (1956), Burr (1960) and Robillard (1972). Brown, Hollander and Korwar (1974) derived tests based on tau for censored data.

For the null distribution of rho see Olds (1938), Kendall, Kendall and Smith (1939), Pitman (1937b), Hotelling and Pabst (1936), Olds (1949), David, Kendall and Stuart (1951), Teegarden (1960) and Nijsse (1988). More extensive tables were developed by Glasser and Winter (1961), Owen (1962), De Jonge and Van Montfort (1972), Zar (1972), Otten (1973a, 1973b), Dunstan, Nix and Reynolds (1979), Neave (1981), Nelson (1986), Franklin (1987a, 1987b) and Ramsay (1988). The latter reference gives the most complete and accurate table of critical values. Franklin (1988c) gave a table of the complete distribution for $n = 12$ to $n = 18$. Iman and Conover (1978) gave approximations to the critical region of rho both with and without the presence of ties. A test based on $\sum d^2$ is sometimes called the Hotelling–Pabst test.

Sen and Krishnaiah (1984) give tables for the null distribution of both tau and rho. Bennett and Choe (1984) studied Fisher's $\tanh^{-1}$ transformation for tests based on rho and tau with small samples.

Kendall and Stuart (1979) indicated that tests of independence based on r_s and t are asymptotically equivalent and have high correlation even for small sample sizes.

For the general joint distribution of correlation coefficients, see Daniels (1944), Daniels and Kendall (1947), Hoeffding (1948a) and Stuart (1956a).

The result in (4.12) for tau was established by Daniels and Kendall (1947), but Daniels (1950) showed that this is a poor upper bound in general. Stuart (1956a) proposed a new bound that is sharper.

The result in (4.13) for rho follows from general results given by Hoeffding (1948a). If k is defined by

$$k = \frac{(n + 1)r_s - 3t}{n - 2},$$

then

$$\mathrm{var}(k) \leq \frac{3(1 - \rho_s^2)}{n},$$

where ρ_s is Spearman's rho for a continuous population (cf. Section 9.5). For large, n, k approaches r_s.

Bell and Doksum (1967) derived lower bounds for the power of general two-sample rank tests of independence.

Kraemer (1985) gave a method for computing the approximate power for tests of independence based on tau or rho. Noether (1987b) gave methods to determine the approximate sample size needed for a test based on tau in order to achieve a prespecified power at a given level of significance; the sample size is a function of the probability of concordance in the population.

Schemper (1987) used Monte Carlo experiments to investigate the performance of one-sample and two-sample tests of Kendall's tau based on the variance estimate given in (4.12), the jackknife variance estimate, the bootstrap variance estimate, and the Edgeworth corrected bootstrap test.

Franklin (1988a) examined the convergence to normality and the accuracy of seven different approximations to the exact null distribution of rho. This paper recommends the tables of Zar (1972) or Pearson Type II approximation for sample sizes greater than 18, and the use of exact tables for smaller sample sizes. Best and Roberts (1975) found an Edgeworth series approximation (not considered by Franklin) superior to both the Pearson and Beta approximations. Ramsay (1989) makes comparisons of the Edgeworth and Pearson approximations and reinforces the conclusion of Best and Roberts; Ramsay gives tables of critical values for $19 \le n \le 100$ based on this Edgeworth approximation. Franklin (1989) compared three different Edgeworth approximations and Pearson approximations. Koroljuk and Borovskih (1982) discuss approximations to the distribution of rho.

Lehmann (1966) discussed concepts of dependence. Joag-Dev (1984) reviewed nonparametric measures of dependence for bivariate distributions and gave references. Kochar and Gupta (1987) proposed a class of distribution-free tests which includes tau for tests of independence against positive quadrant dependence.

Whitfield (1949) proposed an intraclass rank correlation coefficient of the Kendall tau type and gave tables of its sampling distribution. Shirahata (1981) proposed an intraclass correlation coefficient of the Spearman rho type and gave tables of its sampling distribution. Spurrier and Hewett (1980) gave a two-stage test of independence based on Kendall's tau that has application in quality control.

The test for trend based on Kendall's tau was developed by Mann (1945) and the one based on Spearman's rho was suggested by Daniels (1950). Their efficiencies relative to Student's t-test are $(3/\pi)^{1/3}$ for normal distributions as reported by Stuart (1954b, 1956b) and Noether (1958), and at least 0.95 for any distribution. Other nonparametric tests for trend were developed by Cox and Stuart (1955), Moore and Wallis (1943). The efficiencies of all these tests were compared by Aiyar, Guillier and Albers (1979). Bhattacharyya (1984) gave an overview of nonparametric tests for trend or serial correlation. Dietz and Killeen (1981) extend the Mann test for trend to the multivariate case. Regression applications are covered in Chapter 13 of this book.

Problems

4.1 McCaslin and Stanga (1986) studied the extent to which information needs of financial analysts making investment decisions and commercial loan officers making loan decisions are perceived to be the same. Informational needs were assessed from mail questionnaires completed by 56 financial analysts and 113 loan officers in the United States. The respondents rated each of 30 information items on a seven-point Likert scale (1 = least needed) as to accounting (a) relevance and (b) reliability. These 30 information items are divided into three categories called historical cost items, constant dollar items and current cost items. The authors conclude that information needs on all 30 items simultaneously are similar for analysts and bankers, which confirms results of previous studies made in the United Kingdom. The eight current cost items and their mean rankings are shown in Table 4.7 for both relevance and reliability.

Table 4.7

Current cost items	Relevance mean rate		Reliability mean rate	
	analysts	bankers	analysts	bankers
Income from continuing operations	5.351	5.464	4.446	4.634
Cost of goods sold	5.088	5.198	4.386	4.541
Depreciation expense (straight line)	4.737	4.795	4.105	4.288
Depletion expense (units of production)	4.579	4.676	4.053	4.261
Inventory	4.825	5.207	4.125	4.550
Property, plant, equipment	4.772	4.892	3.947	4.393
Net assets at year end	4.667	4.598	4.000	4.152
Earnings per share	5.211	3.333	4.211	3.817

(a)   Investigate the authors' research hypothesis that the mean ratings of bankers and analysts show a positive agreement when rated with respect to

   (i)   relevance
   (ii)   reliability.

(b)   Investigate the research hypothesis that the mean ratings on relevance and reliability show a positive agreement when rated by

   (i)   analysts
   (ii)   bankers.

4.2   Sinclair, Guthrie and Forness (1984) report a study to investigate the claim that attention spans are lower in children with learning disabilities and inattentiveness leads to underachievement in school. Attention was measured in the classroom by observing and recording the behaviour of each child serially in six-second intervals in a round-robin fashion until a minimum of one hour of behaviour had been observed for each child on each of a minimum of four days. Inattentiveness was measured as a percentage to reflect the relative frequency of time when behaviour was observed to be incompatible with task activities for a group of children. Thus a high percentage score on inattentiveness means a low attention span. Actual reading achievement was measured by the WRAT test score (which ranges between 1 and 5 years) and then converted to number of years behind. For the data in Table 4.8 on inattentiveness and years behind in reading achievement for 18 learning disabled children in Los Angeles, test to see whether there is a significant positive relationship between years behind and inattentiveness.

Table 4.8

Years behind	Inattentiveness
1.5	4
4.0	24
3.3	9
2.4	44
1.8	36
4.7	29
3.5	11
4.4	42
2.9	31
1.3	26
2.0	39
4.3	14
3.6	48
3.8	54
2.3	46
3.0	51
1.0	17
5.0	21

Answer: $r_s = 0.063$, P-value > 0.10.

4.3 Bonjean, Brown, Grandjean and Macken (1982) reported a longitudinal study of nursing educators regarding importance of job characteristics and satisfaction with job characteristics in 1979, as compared with 1973, at the University of Texas at Austin. Questionnaires were distributed to faculty administrators and responses were given on a five-point Likert scale (0–4) where 4 = most important. The mean ratings on importance of 21 job characteristics are given in Table 4.9 for all 42 faculty at the school in 1973 and all 50 faculty in 1979 along with the same mean ratings for the 21 persons who were faculty members in both 1973 and 1979 (called stable faculty). Use Spearman's rho to determine whether there is a significant positive agreement between mean responses on importance of job characteristics between 1973 and 1979 for

(a) all faculty.
(b) stable faculty.

Table 4.9

Characteristic	Mean responses			
	All faculty		Stable faculty	
	1973	1979	1973	1979
Opportunity to be a good teacher	3.7	3.8	3.6	3.8
Library facilities	2.6	3.4	2.9	3.1
Supportive colleagues	3.0	3.3	3.3	3.1
Clinical knowledge current	2.9	3.2	2.9	2.9
Dean lets me define my responsibilities	3.4	3.1	3.6	3.1
Unprejudiced evaluations	1.4	3.0	1.5	3.0
Laboratory facilities	2.7	3.0	2.6	3.0
Salary	2.7	2.8	2.9	3.0
Fringe benefits	1.6	2.6	2.1	3.0
Stimulating students	2.1	2.6	2.1	2.4
Time and facilities for research	1.8	2.6	1.8	2.6
Voice in school policy	2.7	2.5	2.8	2.5
Tenure possibilities	1.4	2.4	1.4	2.7
Attractive community	1.2	2.0	1.3	2.0
National recognition of school	1.3	2.0	1.6	2.1
Independence	1.1	2.0	1.3	2.2
Having responsibility	1.3	1.9	1.5	2.2
Physical surroundings at work	1.1	1.9	1.1	2.0
Spouse's career	1.3	1.6	1.1	1.8
Availability of dean	0.8	0.9	1.0	1.6
Minimum of teaching	0.1	0.7	0.2	0.6

Answers: (a) $r_s = 0.89$, *P*-value < 0.001
(b) $r_s = 0.89$, *P*-value < 0.001.

4.4 Duckson (1988) investigated patterns of spatial awareness of college freshmen between 1968 and 1986 using scores on a pre-test given to students in a geography class at Frostburg State College in Maryland. The test was to identify each of the 48 contiguous states on a map of the United States during a 15-minute period. The score was the number of states incorrectly identified, so that a high score indicates poor knowledge of US geography. The first group of subjects was a random sample of 435 students who took the test between 1968 and 1973 and the second group sample was 456 students who took the test between 1980 and 1986. The data given in Table 4.10 as raw scores are the total number of incorrect responses by state for each group. Determine whether there is a significant direct association between Group 1 and Group 2 in the pattern of errors in recognition of the individual states.

4.5 Mitman, Mergandoller, Marchman and Packer (1987) examined the relative amount of time seventh grade life-science teachers spend presenting science content and science context (the role of science in society). Data on 11 randomly selected life-science teachers were

Table 4.10

State	Group 1 Raw score	Group 2 Raw score		State	Group 1 Raw score	Group 2 Raw score
AL	98	170		NE	154	289
AZ	118	215		NV	38	88
AR	190	332		NH	186	299
CA	1	5		NJ	107	173
CO	145	240		NM	60	164
CN	149	249		NY	40	69
DE	63	107		NC	35	51
FL	3	9		ND	34	90
GA	37	64		OH	54	97
ID	80	167		OK	100	191
IL	140	188		OR	45	115
IN	143	184		PA	20	35
IA	185	332		RI	103	196
KS	152	257		SC	38	56
KY	81	159		SD	33	90
LA	88	176		TN	56	100
ME	21	68		TX	2	4
MD	8	22		UT	91	163
MA	112	191		VT	187	251
MI	125	252		VA	21	59
MN	186	293		WA	24	43
MS	124	255		WV	23	47
MO	175	317		WI	193	301
MT	115	201		WY	192	314

obtained for the presentation of two different topics topics during the 1983–84 school year by recording the amount of instructional time in minutes that represents a reference to science content and converting this to a percentage of total instructional time (Table 4.11). Determine whether there is a significant positive relationship between the percentage of time spent on content for the two topics.

Table 4.11

Class	Topic 1	Topic 2
1	86	93
2	100	100
3	99	100
4	95	97
5	100	94
6	99	100
7	99	99
8	100	99
9	94	99
10	98	100
11	95	98

Answer: $r_s = 0.47$, ties are too extensive to report a reasonable P-value.

4.6   Another research hypothesis studied by Mitman *et al.* (1987) was that students' perceived time spent on science content is directly related to students' perceived time spent on context. The data here came from a survey of students to evaluate the extent to which their teacher addressed content and context (1 = never, 5 = very often). The data in Table 4.12 are the average class scores of students for content and context. Is the research hypothesis supported by these data?

Table 4.12

Class	Content score	Context score
1	4.14	3.33
2	3.83	2.79
3	3.81	2.61
4	4.51	3.29
5	3.65	2.64
6	4.06	3.10
7	3.67	2.82
8	4.42	2.89
9	3.78	2.81
10	4.15	3.02
11	3.82	2.74

Answers: $r_s = 0.718$, *P*-value = 0.0081
$t = 0.491$, *P*-value = 0.025.

4.7   One further research hypothesis investigated by Mitman *et al.* (1987) was that students' perceived time spent on context is directly related to teachers' actual time spent on context. Use the data below to determine whether this claim is substantiated.

Table 4.13

Class	Teachers' context time	Students' context score
1	11.0	3.33
2	0.0	2.79
3	0.2	2.61
4	3.8	3.29
5	4.0	2.64
6	0.7	3.10
7	0.5	2.82
8	0.5	2.89
9	4.0	2.81
10	1.2	3.02
11	3.5	2.74

Answer: $r_s = 0.34$, $0.150 < $ *P*-value $ < 0.157$.

4.8 Ashton (1985) examined the relationship between consistency of decision makers and accuracy of decision makers in accounting literature. This study is of interest because decision makers frequently use consistency as a proxy for accuracy because actual accuracy figures are not available. Consistency is measured by the correlation between predictions by each pair of subjects and accuracy is measured by the mean accuracy of each pair of subjects, where individual accuracy is measured by the correlation between the actual outcome and each subject's prediction. Determine whether consistency is a good proxy for accuracy by seeing if their rank correlation is highly significant for the data in Table 4.14 on 13 subjects.

Table 4.14

Subject	Accuracy	Consistency
1	0.879	0.760
2	0.650	0.472
3	0.824	0.715
4	0.746	0.614
5	0.678	0.596
6	0.691	0.653
7	0.916	0.782
8	0.619	0.522
9	0.689	0.556
10	0.607	0.451
11	0.898	0.763
12	0.614	0.516
13	0.857	0.746

4.9 Cancian and Ross (1981) studied the changes in media coverage of women over the period of the women's movement in the United States. One aspect of the study reported the number of items about women covered on the TV evening news by all three major networks combined between 1968 and 1977 (Table 4.15). Test to see if there is a significant positive trend.

Table 4.15

Year	Number of items	Year	Number of items
1968	22	1973	58
1969	18	1974	97
1970	53	1975	192
1971	41	1976	111
1972	68	1977	194

4.10 Whipple and Niedell (1971–72) studied attitudes of blacks and whites toward competing non-food stores in Buffalo, New York. Data were obtained as overall rankings of ten stores with respect to favourableness of image of the stores (1 = best) and number of store visits (1 = most) (Table 4.16). Use Spearman's rho to investigate the research hypothesis that there is a positive correlation between

(a)   favourableness and store visits for blacks;
(b)   favourableness and store visits for whites;
(c)   black favourableness and white favourableness;
(d)   black store visits and white store visits.

Table 4.16

Store	Blacks		Whites	
	Favourableness	Visits	Favourableness	Visits
AM & A	1	1	1	1
GEX	9	9	4	10
Grant	3	7	7	4
Hengerer	2	6	3	7.5
Hens & Kelly	4	8	2	5
Sears	7	4	5	3
IDS	6	5	6	2
Twin Fair	8	3	10	6
Two Guys	10	10	8	9
Sattler's	5	2	9	7.5

Answers: (a) $\sum d^2 = 92$, *P*-value 0.102
(b) $\sum d^2 = 113.5$, $0.184 < P\text{-value} < 0.193$
(c) $\sum d^2 = 74$, *P*-value 0.052
(d) $\sum d^2 = 71.5$, $0.044 < P\text{-value} < 0.048$

4.11   Reierson (1966) reported a survey of attitudes of US college students at Baylor University and Texas A&M toward foreign products. As one part of the study overall rankings of 26 specific foreign products by 105 business students at Baylor and 50 psychology students at Texas A&M were compared by computing Spearman's rank correlation coefficient. For the data in Table 4.17, verify their result that $r_s = 0.856$, compute Kendall's tau, and investigate the research hypothesis of direct association between attitudes of students at Baylor and Texas A&M.

4.12   Ingram (1984) investigated the association between an index of quantity of US state government accounting disclosure practices and each of 14 independent variables selected as surrogates to measure political and economic factors in the state for all 50 states. For example, the surrogate for state wealth was state revenue *per capita*. The resulting correlations are shown in Table 4.18. Investigate the research hypothesis that there is a positive relationship between the Pearson and Spearman correlation coefficients.

Table 4.17

Product	Baylor rank	A&M rank
English		
Furniture	5	5
Washing Machine	21	21
Soft Drinks	18	20
Suits	2	3
Automobiles	7	8
Candy	13	13
Japanese		
Shoes	26	26
Television Sets	20	22
Candy	24	24
China	16	9
Cameras	12	12
Rice	8	7
Italian		
Furniture	10	15
Refrigerators	25	25
Soft Drinks	23	23
Cheese	4	6
Shoes	9	17
Automobiles	15	16
Spaghetti	1	1
Office Machines	19	19
Swedish		
Radios	11	10
Candy	6	11
Sweaters	3	4
Automobiles	17	2
Soft Drinks	22	18
Furniture	14	14

Answers: $r_s = 0.856$, *P*-value < 0.001
$t = 0.729$, *P*-value < 0.005.

Table 4.18

Variable	Pearson	Spearman
Political competition	0.187	0.182
Urbanisation	0.205	0.296
Per capita income	0.349	0.272
Median school years	0.149	0.106
Appointive powers	0.275	0.301
Selection of accounting administrator	0.234	0.256
Selection of auditor	0.291	0.296
Newspaper circulation	−0.012	−0.004
Long-term debt *per capita*	0.160	0.138
Intergovernmental revenue	0.023	0.157
Revenue *per capita*	0.322	0.315
Governor's salary	0.205	0.306
Legislators' salary	0.207	0.124
Accounting salaries	0.283	0.362
Auditor-CPA code	0.146	0.122
Population	0.126	0.131

4.13 Holden and Holden (1978) evaluated the relationship between each of effective tariff rate (ETR) and nominal tariff rate (NTR) with five variables for a sample of 28 industries in South Africa. The variables used were output (Y), percentage change over time in output (ΔY), percentage change in employment (ΔE), ratio of imports to domestic supply (I/S) and percentage change in I/S $(\Delta(I/S))$. Both the Pearson and Spearman rank correlation coefficients were computed. Investigate the research hypothesis that there is a positive relationship between the Pearson and Spearman coefficients for the five variables in the case of each of effective tariff rates and nominal tariff rates for the data in Table 4.19.

Table 4.19

Variable	ETR		NTR	
	Pearson	Spearman	Pearson	Spearman
Y	−0.09	0.03	−0.11	−0.07
ΔY	−0.17	−0.19	−0.10	0.05
ΔE	−0.28	−0.20	−0.23	−0.14
I/S	−0.16	−0.24	−0.15	−0.16
$\Delta (I/S)$	−0.11	−0.13	−0.11	−0.11

4.14 Ezell and Ward (1983) compared the attitudes between frequent and less frequent patrons of beauty salons with respect to 12 factors influencing their choice of which salon to patronise. Respondents were 150 less frequent working woman patrons (defined as less than once a month) and 141 frequent working woman patrons. Each woman was asked to rank the factors from 1 to 12, with 1 indicating most important. Determine whether there is a significant positive relationship between attitudes of frequent and less frequent patrons using the median ranks shown in Table 4.20.

Table 4.20

Factor	Frequent patrons Median value	Less-frequent patrons Median value
Quality of work	1.327	1.245
Cleanliness	2.500	4.958
Flexibility in scheduling appointments	4.125	7.031
Cost	4.714	4.786
Adherence to schedule	4.812	7.346
Friendliness of operator	5.143	4.969
Operator's knowledge of style trends	5.167	4.929
Convenience of location	5.714	6.550
Hours of operation	5.769	6.667
Shop appearance and atmosphere	6.500	7.833
Variety of services offered	9.167	10.766
Others' recommendations	10.531	8.182

4.15 Peters (1987) gave some figures obtained from The Gallup Organization that compare the percentage of popular vote predicted for the winner by the final Gallup survey and the percentage actually obtained by the winner for the US presidential elections between 1936 and 1984 (Table 4.21). Test to see if there is a significant positive agreement between the rankings of these percentages.

Table 4.21

Year	President	Prediction	Vote
1936	Roosevelt	55.7	62.5
1940	Roosevelt	52.0	55.0
1944	Roosevelt	51.5	53.3
1948	Truman	44.5	49.9
1952	Eisenhower	51.0	55.4
1956	Eisenhower	59.5	57.8
1960	Kennedy	51.0	50.1
1964	Johnson	64.0	61.3
1968	Nixon	43.0	43.5
1972	Nixon	62.0	61.8
1976	Carter	48.0	50.0
1980	Reagan	47.0	50.8
1984	Reagan	59.0	59.2

4.16 McDaniel and Hise (1984) obtained information from 236 chief executive officers (CEO) of US industrial corporations about their perception of the importance of eleven marketing activities and five nonmarketing activities in their firm (Table 4.22). Customer relations

Table 4.22

	Marketing background	Nonmarketing background
Marketing activity		
Advertising	2.73	2.24
Customer relations	4.31	4.25
Marketing analysis	3.81	3.49
Marketing research	3.62	2.17
New product development	4.11	3.69
Personal selling	3.47	3.00
Physical distribution	3.14	3.24
Pricing strategies	3.81	3.88
Product management	3.75	3.55
Public relations	2.83	2.71
Sales promotion	3.36	3.10
Nonmarketing activity		
Financial planning	4.33	4.37
Labour relations	3.80	3.72
Personnel management	3.81	3.66
Production manufacturing	3.98	4.16
Research and development	3.88	3.58

was ranked most important and advertising was least important among marketing activities. Financial planning was most important and research and development was least important among nonmarketing activities. The authors then separated the CEOs according to whether their primary work background was marketing or nonmarketing and presented the mean ratings for each group (5 = very important). Test for CEO rating agreement between groups on

(a)   marketing activities
(b)   nonmarketing activities.

Chapter 5

Proof of the results of Chapter 4

5.1 The series of formulae we have used in the last chapter for testing the significance of t and r_s require the following four types of result:

(a) the derivation of exact distributions for small values of n;
(b) the proof that distributions tend to normality for large n;
(c) the derivation of the means and variances of the limiting distributions;
(d) the derivation of the corrections for continuity.

We treat them in that order and conclude the chapter with a more detailed analysis of the non-null case.

Exact distribution of t in the null case

5.2 If we correlate a fixed ranking of n members with the $n!$ possible rankings (excluding ties) we obtain the same distribution whatever the fixed ranking; for all possibilities occur. We therefore lose no generality in supposing our fixed ranking to be the natural order $1, \ldots, n$. Let $u(n, S)$ be the number of values of S in the aggregate of $n!$ possible values obtained by correlating this with all possible rankings.

Consider one such ranking and the effect of inserting a new member $(n + 1)$ at the various places in it, from the first (preceding the first member p_1) to the last (following the last member p_n). Inserting $(n + 1)$ at the beginning will add $-n$ to S; inserting it at the second place adds $-(n - 2)$; at the third place $-(n - 4)$ and so on, the last place adding n. It follows that

$$u(n + 1, S) = u(n, S - n) + u(n, S - n + 2) + u(n, S - n + 4) + \cdots$$
$$+ u(n, S + n - 4) + u(n, S + n - 2) + u(n, S + n). \tag{5.1}$$

This recurrence formula enables us to ascertain the frequency-distribution of S for $n + 1$ when we know that for n; and hence we may build up the series of distributions from the simpler cases for $n = 2, 3$, etc.

5.3 In practice the procedure may be simplified. For $n = 2$ there are two values of S, namely -1 and $+1$. If we write the frequencies down three times, one under the other, moving a stage to the right each time, we get

1	1		
	1	1	
		1	1
1	2	2	1

and the sum gives the frequencies of S for $n = 3$, the actual values ranging from $-\frac{1}{2}n(n - 1)$ by units of two to $\frac{1}{2}n(n - 1)$, i.e. through the values -3, -1, $+1$, $+3$.

Similarly for $n = 4$ we write down the array for $n = 3$ four times and sum, as follows:

1	2	2	1				
	1	2	2	1			
		1	2	2	1		
			1	2	2	1	
1	3	5	6	5	3	1	

the frequencies being for values of S from -6 by units of 2 to $+6$.

The validity of this rule will be evident from the previous section. To any set of S-values for a given n, there will correspond in the distribution for $n + 1$ a similar set with values of S increased by $-n$, $-(n - 2)$, $\ldots$, $(n - 2)$, n. What we have done is to write these frequencies down, one row for $-n$, one for $-(n - 2)$ and so on, and to sum them.

5.4 Even this procedure may be simplified. We form a numerical triangle as follows:

n				Frequencies						
1	1									
2	1	1								
3	1	2	2	1						
4	1	3	5	6	5	3	1			
5	1	4	9	15	20	22	20	15	9	4	1
etc.										

In this array a number in the rth row is the sum of the number immediately above it and the $(r - 1)$ members to the left of that number: e.g. in the fifth row $22 = 3 + 5 + 6 + 5 + 3$. The frequencies of values of S for $n = r$ are those so obtained in the rth row. This method was used to obtain the frequencies and therefore the probabilities for $n \leq 10$ in Appendix Table 1.

5.5 The distribution of S is symmetrical; for to any ranking giving a particular value of S there will correspond a conjugate ranking giving $-S$. The mean value of S is therefore zero, as we should expect. That S ranges from $-\frac{1}{2}n(n - 1)$ to $\frac{1}{2}n(n - 1)$ is evident from the fact that these are the extreme values given when the ranking is the inverse of the natural order or the natural order itself. Further, the interval between successive values of S for given n is 2. This follows from (5.1), or perhaps more simply by the consideration that an interchange of a pair of members in a ranking alters the positive score by $+1$ or -1 and hence the difference between positive and negative scores (which is S) by $+2$ or -2.

Tendency of *t* to normality in the null case

5.6 It remains to determine the variance of S and to prove that the distribution tends to normality as n increases.

In the notation of Chapter 2 let us write

$$c_{ij} = a_{ij}b_{ij}, \tag{5.2}$$

where a and b refer to the scores in the different rankings. Define

$$c = \sum_{i,j=1}^{n} c_{ij} \tag{5.3}$$

so that

$$c = 2S. \tag{5.4}$$

Now

$$\sum_{l=1}^{n} a_{il} = n + 1 - 2i \tag{5.5}$$

since this is the score associated with the ith member and is $(n-i) - (i-1)$. Further,

$$\sum_{i,l=1}^{n} a_{il}^{2} = n(n-1) \tag{5.6}$$

for a^2 is $+1$ and this sum is merely the number of possible ways of choosing a pair from n members, each pair being counted twice, once as AB and once as BA. It follows from (5.5) that

$$\sum_{i,l=1}^{n} a_{il} = \sum_{i=1}^{n} (n + 1 - 2i)$$

$$= n(n + 1) - 2 \sum_{i=1}^{n} i$$

$$= 0,$$

as is to be expected.

We also find

$$\sum_{i,l,t=1}^{n} a_{il} a_{it} = \sum_{i,l} a_{il}(n + 1 - 2i)$$

$$= \sum_{i} (n + 1 - 2i)^2$$

$$= \sum_{i} (n + 1)^2 - 4(n + 1) \sum i + 4 \sum i^2$$

$$= n(n + 1)^2 - 2n(n + 1)^2 + \tfrac{2}{3}n(n + 1)(2n + 1)$$

$$= \tfrac{1}{3}n(n^2 - 1). \tag{5.7}$$

Now writing E to denote mean values on summation over all possible permutations, we have

$$E(c) = E \sum_{i,j=1}^{n} (a_{ij} b_{ij})$$

$$= \sum_{i,j} E(a_{ij} b_{ij}).$$

Since a_{ij} and b_{ij} are independent, any fixed value of one being taken with all possible values of the other, and since the mean value E of any term a_{ij} or b_{ij} is zero, we have

$$E(c) = 0, \tag{5.8}$$

confirming that the mean of c (or of S) is zero. For the variance of c we require

$$E(c^2) = E\left\{ \sum_{i,j} (a_{ij} b_{ij}) \right\}^2$$

$$= E\{ \sum (a_{ij}^2 b_{ij}^2) + \sum{}' (a_{ij} b_{ij} a_{ik} b_{ik}) + \sum{}'' (a_{ij} b_{ij} a_{kl} b_{kl}) \} \tag{5.9}$$

where $i \neq j \neq k = l$. The summations here take place over the terms specified, but any term may arise in more than one way from the expansion of $(\sum a_{ij} b_{ij})^2$.

(i)   The term $E \sum{}''$ vanishes. To demonstrate this it is enough to show that $E \sum{}'' a_{ij} a_{kl}$ vanishes. Now ($\sum$ relating to summation over all values of the suffixes)

$$\sum a_{ij} a_{kl} = \sum{}'' a_{ij} a_{kl} + \sum{}' a_{ij} a_{il} + \sum{}' a_{ij} a_{ki} + \sum{}' a_{ij} a_{jl}$$

$$+ \sum{}' a_{ij} a_{kj} + \sum a_{ij} a_{ij} + \sum a_{ij} a_{ji}.$$

On taking expectations, the terms after the first on the right vanish in pairs in virtue of such relations as $a_{ij} = -a_{ji}$. The term on the left vanishes because (summation extending over all values) $E \sum a_{ij} a_{kl} = E \sum a_{ij} E \sum a_{kl}$ and $\sum a_{ij} = 0$. Consequently the first term on the right vanishes.

(ii)   Now consider the sum $E \sum a_{ij}^2 b_{ij}^2$. An individual term arises from the square of $a_{ji} b_{ij}$ or the product of that term with $a_{ij} b_{ji}$. Thus the sum is twice the sum of $a_{ij}^2 b_{ij}^2$ with i, j from 1 to n. There are $n(n-1)$ terms in the sum, and the expectation is thus

$$2n(n-1)E(a_{12}^2 b_{12}^2) = 2n(n-1)E(a_{12}^2)E(b_{12}^2) = 2 \sum a_{ij}^2 \sum b_{ij}^2 / n(n-1).$$

(iii)   Similarly the second term on the right in (5.9) is

$$\frac{4}{n(n-1)(n-2)} \sum{}' a_{ij} a_{ik} \sum{}' b_{ij} b_{ik}.$$

For the product $a_{ij} a_{ik} b_{ij} b_{ik}$ can arise in four ways, since if we fix i, j, k, for the a's the b terms can appear with suffixes (ij, ik), (ji, ik), (ij, ki) and (ji, ki); and there are $n(n-1)(n-2)$ ways of fixing three different suffixes out of n.

The reader who has trouble with these factors is advised to write out in full the case for $n = 4$.

Furthermore

$$\sum' a_{ij} a_{ik} = \sum a_{ij} a_{ik} - \sum a_{ij}^2.$$

Substituting in (5.9) we thus find

$$E(c^2) = \frac{4}{n(n-1)(n-2)} \{\sum a_{ij} a_{ik} - \sum a_{ij}^2\}\{\sum b_{ij} b_{ik} - \sum b_{ij}^2\}$$

$$+ \frac{2}{n(n-1)} \sum a_{ij}^2 \sum b_{ij}^2. \tag{5.10}$$

Now substituting from (5.6) and (5.7) we find

$$E(c^2) = \frac{4}{n(n-1)(n-2)} \{\tfrac{1}{3}n(n^2 - 1) - n(n-1)\}^2 + \frac{2}{n(n-1)} [n(n-1)]^2$$

$$= \frac{2n(n-1)(2n+5)}{9}. \tag{5.11}$$

It follows from (5.4) that

$$E(S^2) = \text{var } S = \frac{n(n-1)(2n+5)}{18} \tag{5.12}$$

as given in (4.2), and from (1.3),

$$\text{var } t = \frac{2(2n+5)}{9n(n-1)}. \tag{5.13}$$

5.7   If ties are present, equation (5.10) remains valid but expressions (5.6) and (5.7) require modification. In place of (5.6) we have

$$\sum a_{ij}^2 = n(n-1) - \sum u(u-1) \tag{5.14}$$

where summation $\sum$ takes place over the various ties. This follows simply from the consideration that for a pair of tied ranks $a_{ij} = 0$ and consequently the sum of the squares of contributions from a tied set is of the same form as for the ranking as a whole.

In place of (5.7) we have

$$\sum a_{ij} a_{ik} = \tfrac{1}{3}n(n^2 - 1) - \tfrac{1}{3}\sum u(u^2 - 1). \tag{5.15}$$

This is not quite so obvious. Consider the effect of tying a set of ranks. The contribution to the sum on the left of (5.15) will be unchanged if the suffix i falls outside this set. If i is inside the set and j, k are outside, again the contribution is unchanged. If both j, k fall inside no contribution arises, and therefore we have to substract the term $\tfrac{1}{3}u(u^2 - 1)$. If one falls inside and one outside the contribution remains unchanged, for it was zero in the original untied case, each possible pair occurring once to give $+1$ and once to give -1.

If we substitute from (5.14) and (5.15) in (5.10) we find, for the variance of S when the rankings contain ties typified by u, v,

$$\text{var } S = \tfrac{1}{18}[n(n-1)(2n+5) - \sum u(u-1)(2u+5) - \sum v(v-1)(2v+5)]$$

$$+ \frac{1}{9n(n-1)(n-2)}[\sum u(u-1)(u-2)][\sum v(v-1)(v-2)]$$

$$+ \frac{1}{2n(n-1)}[\sum u(u-1)][\sum v(v-1)]. \tag{5.16}$$

This is the result given in (4.3). Equations (4.4), (4.5) and (4.6) follow at once.

5.8   In proving that the distribution of S tends to normality we shall follow a procedure which, with a few trivial changes, will also prove the normality of $\sum d^2$ and will form an introduction to more general results, due to Daniels (1944), concerning the limiting forms of the general rank correlation coefficients as defined in Chapter 2.

We prove that the moments of the distribution of S tend to those of the normal distribution. It will then follow from what is known as the Second Limit Theorem that the distribution tends to normality. Since the distribution of S is symmetrical moments about the mean of odd order vanish. We have only to show that for the even moments

$$\mu_{2r} = \frac{(2r)!}{2^r r!}(\mu_2)^r. \tag{5.17}$$

Consider the mean value of $(\sum a_{ij} b_{ij})^{2r}$. When this expression is expanded it will consist of terms such as

$$\sum{}' a_{ij} a_{kl} a_{mn} \cdots b_{ij} b_{kl} b_{mn} \cdots.$$

When summations extend over values of suffixes which exclude certain values (e.g. $i \neq k$) we can replace them by complete summations. Our expanded expression will then, apart from numerical terms, consist of terms such as

$$\sum a_{ij} a_{kl} a_{mn} \cdots \sum b_{ij} b_{kl} b_{mn} \cdots, \tag{5.18}$$

where certain suffixes may be the same or 'tied' and others will be different and 'free'. Consider a term in which the $4r$ suffixes of a are tied in pairs, e.g.

$$\sum a_{ij} a_{ik} a_{lm} a_{ln} \cdots. \tag{5.19}$$

There are then $3r$ independent suffixes. Now $\sum a_{ij} a_{ik}$ is of order $\tfrac{1}{3}n^3$ and consequently (5.19) is of order $(\tfrac{1}{3}n^3)^r$. In the expansion of $(\sum a_{ij} b_{ij})^{2r}$ terms of this type will arise with a frequency which is the number of ways of tying the $2r$ paired suffixes. This is the product of three factors, namely (i) the number of ways of picking r *a*s from $2r$, namely $\binom{2r}{r}$; (ii) the number of ways of associating these with the remaining r factors, namely $r!$; (iii) 2^r arising from the fact that either suffix of a particular a may be tied. The numerical coefficient is then

$$\binom{2r}{r} r!\, 2^r = \frac{(2r)!\, 2^r}{r!}.$$

Furthermore, if $3r$ suffixes are fixed the remaining members of the ranking can vary in $(n - 3r)!$ ways. Hence μ_{2r} contains a term

$$\frac{(n - 3r)!}{n!} \frac{2^r (2r)!}{r!} (\tfrac{1}{3} n^3)^{2r} \sim \frac{(2r)!}{2^r r!} \left(\frac{4}{9} n^3\right)^r.$$

But μ_2 is of order $\tfrac{4}{9} n^3$ (as we see by putting $r = 1$) and hence μ_{2r} contains a term

$$\frac{(2r)!}{2^r r!} (\mu_2)^r.$$

Our demonstration will be complete if we show that all other terms are of lower order in n.

If a term in (5.18) contains a pair of suffixes neither of which appears elsewhere, the term vanishes because $\sum a_{ij} = 0$. Consider a term where more than two suffixes are tied. The summation of (5.19) is then over not more than $3r - 1$ suffixes and cannot be of greater order than $(n^{3r-1})^2$. If $3r - 1$ or fewer suffixes are fixed the order is then not greater than

$$\frac{(n - 3r + 1)!}{n!} n^{6r-2} \sim n^{3r-1}$$

and is thus lower than that of the term already found.

Distribution of r_s in the null case

5.9   The actual distribution of r_s is much more difficult to determine than that of t. Consider the array

$$
\begin{array}{cccccccc}
0 & 1 & 2 & \cdots & n-3 & n-2 & n-1 \\
-1 & 0 & 1 & \cdots & n-4 & n-3 & n-2 \\
-2 & -1 & 0 & \cdots & n-5 & n-4 & n-3 \\
\vdots & \vdots & \vdots & & \vdots & \vdots & \vdots \\
-(n-2) & -(n-3) & -(n-4) & \cdots & -1 & 0 & 1 \\
-(n-1) & -(n-2) & -(n-3) & \cdots & -2 & -1 & 0
\end{array}
\qquad (5.20)
$$

Any permissible set of deviations between the order $1, 2, \ldots, n$ and an arbitrary order is given by selecting a member from the array so that no two members appear in more than one row or column. Consider then the array

$$
\left\{
\begin{array}{cccccccc}
a^0 & a^1 & a^4 & \cdots & a^{(n-3)^2} & a^{(n-2)^2} & a^{(n-1)^2} \\
a^1 & a^0 & a^1 & \cdots & a^{(n-4)^2} & a^{(n-3)^2} & a^{(n-2)^2} \\
\vdots & \vdots & \vdots & & \vdots & \vdots & \vdots \\
a^{(n-2)^2} & a^{(n-3)^2} & a^{(n-4)^2} & \cdots & a^1 & a^0 & a^1 \\
a^{(n-1)^2} & a^{(n-2)^2} & a^{(n-3)^2} & \cdots & a^4 & a^1 & a^0
\end{array}
\right\}
\qquad (5.21)
$$

The indices in (5.21) are the squares of the entries in (5.20), and if we expand (5.21) we shall get the totality of values of $\sum d^2$. By 'expanding' we mean developing the array by selecting, in the $n!$ possible ways, n factors from (5.21) such that no two have a row or column in common, multiplying them for each set of factors and summing the $n!$ resultants. This method was used to arive at the distributions which form the basis of the probabilities for $n \leq 13$ in Appendix Table 2.

5.10   Certain simple properties of the distribution of $\sum d^2$ are evident from elementary considerations. First, any value of $\sum d^2$ must be even; for $\sum d = 0$ and hence the number of odd values of d, and thus d^2, must be even. Secondly, the distribution is symmetrical, for to any value of $\sum d^2$ there corresponds a value $\frac{1}{3}(n^3 - n) - \sum d^2$ derived from the conjugate ranking (Section 2.9). Thirdly, the mean of the distribution is $\frac{1}{6}(n^3 - n)$. Fourthly, the values of $\sum d^2$ range from 0 to $\frac{1}{3}(n^3 - n)$.

5.11   We may determine the variance of $\sum d^2$ in the manner of Section 5.6. From the way in which we derived (5.10) in terms of the scores a and b it will be clear that the equation is equally true for c when the scores relate to Spearman's r_s, for without loss of generality we may write

$$a'_{ij} = j - i, \qquad (5.22)$$

where we have added a prime to distinguish it from the score where t is concerned. We have at once

$$\sum_{j=1}^{n} a'_{ij} = \tfrac{1}{2}n(n + 1 - 2i) \qquad (5.23)$$

and

$$\sum_{i,j=1}^{n} a'^2_{ij} = \sum (j - i)^2$$

$$= 2n \sum_{j=1}^{n} j^2 - 2\left(\sum_{j=1}^{n} j \right)^2$$

$$= \tfrac{1}{6}n^2(n^2 - 1)$$

and

$$\sum_{i,j,k=1}^{n} a'_{ij} a'_{ik} = \tfrac{1}{12}n^3(n^2 - 1). \qquad (5.24)$$

Substitution in (5.10) now gives

$$E(c^2) = \frac{n^4(n - 1)(n + 1)^2}{36}. \qquad (5.25)$$

Remembering that

$$r_s = 1 - \frac{6 \sum d^2}{n^3 - n}$$

and that from (2.9)

$$c = \tfrac{1}{6}n^2(n^2 - 1) - n \sum d^2,$$

we find from (5.25)

$$\text{var } r_s = E(r_s) = \frac{1}{n - 1},\qquad(5.26)$$

which is the formula (4.8).*

5.12   By similar methods it may be shown that the fourth moment of r_s is given by

$$\mu_4 = \frac{3(25n^3 - 38n^2 - 35n + 72)}{25n(n + 1)(n - 1)^3}.\qquad(5.27)$$

The third moment, of course, is zero because of the symmetry of the distribution. Expressions are known for μ_6 and μ_8 (see David, Kendall and Stuart, 1951), but they are rather complicated.

In a normal distribution

$$\mu_4 - 3\mu_2^2 = 0.\qquad(5.28)$$

For the distribution of p we find, from (5.27) and (5.26),

$$\mu_4 - 3\mu_2^2 = -\frac{114n^2 + 30n - 216}{25n(n + 1)(n - 1)^3}\qquad(5.29)$$

$$= -\frac{4.56}{n^3} + 0(n^{-4}).\qquad(5.30)$$

This is of lower order than μ_2^2 and equivalently we have

$$\frac{\mu_4}{\mu_2^2} - 3 = -\frac{4.56}{n} + 0(n^{-2}),\qquad(5.31)$$

showing that for large n the distribution of r_s is nearly normal.

For most ordinary work the tables of exact values together with the normal approximation are sufficient. For more precise work David, Kendall and Stuart (1951) obtained expansions giving the probability distribution of both r_s and t to about four-figure accuracy.

Joint distribution of t and r_s

5.13   The tendency of r_s to normality may be demonstrated by making a few alterations in the proof of the normality of t in Section 5.8. Both results

---

*That this formula requires no modification for tied ranks follows from (7.14) with $m = 2$, bearing in mind that (6.6) is to be modified in the tied case. Alternatively, with c as defined in (5.3) and scores $x_j - x_i$, $y_j - y_i$, it may be shown that

$$E(c^2) = 4n^2 \sum (x - \bar{x})^2 \sum (y - \bar{y})^2/(n - 1),$$

and (5.26) follows whether ranks are tied or not.

are particular cases of a more general theorem due to Daniels (1944) which we now prove.

We shall show, in fact, that the joint distribution of t and r_s tends to the bivariate normal form as n tends to infinity. Indeed, under certain non-restrictive conditions, any two coefficients of the general type defined in Section 2.2 tend to joint bivariate normality.

Suppose that a, a' refer to scores in two such coefficients, and similarly for b and b'. Then we shall show that the product-moments of the joint distribution of the corresponding c and c' tend to those of the normal bivariate form.

The pth order product-moments of this joint distribution are sums of terms containing

$$\sum{}' a_{gh} a_{ij} a_{kl} \cdots \sum{}' b_{rs} b_{tu} b_{vw} \cdots, \tag{5.32}$$

where groups of suffixes within the $\sum$'s may be tied or free. Each $\sum'$ involves products of p scores which may belong to either system. Every such $\sum'$ is in turn a linear combination of the corresponding $\sum$ having the same suffixes and other $\sum$s in which additional tied suffixes appear. No $\sum$ may contain a pair of free suffixes attached to one score, for it would then vanish by virtue of the fact that $\sum a_{ij} = 0$.

We discuss first the moments of even order. Let $p = 2m$. Consider a $\sum$ in which the $2m$ scores are divided into m pairs each having one tied suffix, so that there are in all $3m$ independent suffixes, e.g.

$$\sum a_{ij} a_{ik} a_{lr} a_{tu} a_{tv} \cdots. \tag{5.33}$$

It may be written as

$$\left(\sum a_{ij} a_{ik} \right)^{\lambda} \left(\sum a_{ij} a'_{ik} \right)^{\mu} \left(\sum a'_{ij} a'_{ik} \right)^{\nu}, \tag{5.34}$$

where $\lambda + \mu + \nu = m$ and λ, μ, ν are the number of times the scores are paired in the combinations indicated.

As is always possible, suppose the numerically largest value of a_{ij} to be made equal to unity. We now impose the condition that $\sum a_{ij} a_{ik}$ is of order n^3, whether a_{ij} and a_{ik} belong to the same or different systems of scores. This, in particular, is satisfied by t and r_s, when $\max a_{ij} = 1$. With this condition it is seen that $\sum$s of the above types are of order n^{3m}.

Other ways of tying suffixes give $\sum$s of lower order of magnitude. For the order of magnitude of the expression is not reduced on replacing each a_{ij} by $+1$; consequently if further suffixes are tied the order of $\sum$ is made less than n^{3m} since there are fewer than $3m$ summations from 1 to n. It follows that the dominant term in a $\sum'$ is the corresponding $\sum$ having the same array of suffixes.

Moreover, every non-vanishing $\sum$ involving $3m$ independent suffixes can only be a permutation of type (5.32), while those with more than $3m$ different suffixes must all vanish. This will be clear by considering how the $3m$ suffixes can be arrayed between the $2m$ scores. Begin by assigning $3m$ different suffixes at random among the $4m$ available places. At least m scores will receive their full complement of suffixes, all of which will be different. There cannot be more than m such completed scores, for if Σ is not to vanish at least one suffix of each complete pair must be tied, and this can only be done by repeating one suffix from every complete pair in each of the remaining places to be filled, of which there are only m. We are thus led to a permutation of the type of Σ discussed above. If there had been more than $3m$ different suffixes to begin with, there would not have remained sufficient empty places to prevent the existence of at least one score with a pair of free suffixes. Hence all Σs with more than $3m$ different suffixes must vanish.

Any $2m$th product-moment is the sum of terms like

$$\frac{(n-f)!}{n!} A \sum' a_{ij} a_{kl} \cdots \sum' b_{rs} b_{tu} \cdots,$$

where f is the number of independent suffixes in the Σ's and A is a coefficient which is of unit order so far as n is concerned. From the preceding argument the maximum value of f is $3m$, in which case the term is of order

$$n^{-3m} n^{3m} n^{3m} = n^{3m}.$$

When $f \le 3m - 1$ the term is of order not greater than

$$n^{-3m+1}(n^{3m-1})^2 = n^{3m-1}$$

and hence such terms may be neglected. Write

$$\begin{aligned}
h_{11} &= \sum a_{ij} a_{ik} \sum b_{tu} b_{tv} \\
h_{12} &= \sum a_{ij} a'_{ik} \sum b_{tu} b'_{tv} \\
h_{22} &= \sum a'_{ij} a'_{ik} \sum b'_{tu} b'_{tv}.
\end{aligned} \qquad (5.35)$$

Then if lower-order terms are neglected, the even product-moment μ_{rs}, $r + s = 2m$ is given by the sum of terms like

$$n^{-3m} A_{\lambda\mu\nu} h_{11}^{\lambda} h_{12}^{\mu} h_{22}^{\nu}, \qquad 2\lambda + \mu = r, \quad \mu + 2\nu = s \qquad (5.36)$$

over all possible values of λ, μ, ν. The coefficient $A_{\lambda\mu\nu}$ is determined by the following argument. Consider a Σ whose array of suffixes is such that it can be factor as $(\sum a_{ij} a_{ik})^{\lambda} (\sum a_{ij} a'_{ik})^{\mu} (\sum a'_{ij} a'_{ik})^{\nu}$. Its suffix pairs can be permuted in $r! \, s!$ ways within the sets of scores of the two types a and a', but of these $\lambda! \, (2!)^{\lambda} \mu! \, \nu! \, (2!)^{\nu}$ give essentially the same Σ. The suffixes within pairs attached to each score may also be rearranged in 2^{2m} ways without affecting the result. Hence

$$A_{\lambda\mu\nu} = \frac{r! \, s! \, 2^{2m}}{\lambda! \, \mu! \, \nu! \, 2^{\lambda+\nu}} = \frac{r! \, s! \, 2^{m+\mu}}{\lambda! \, \mu! \, \nu!}. \qquad (5.37)$$

From (5.35), (5.36) and (5.37) it follows that the calculation of μ_{rs} is tantamount to determining the coefficient of $t_1^r t_2^s$ in

$$\frac{2^m}{n^{3m} \, m!} (h_{11} t_1^2 + 2h_{12} t_1 t_2 + h_{22} t_2^2)^m. \qquad (5.38)$$

We now consider the odd moments. For r_s and t these vanish by symmetry, but even in the more general case it can be shown that they are negligible to order $n^{-1/2}$. For a $\sum$ containing $2m + 1$ scores cannot have more than $3m + 1$ suffixes. This follows by a similar argument to that employed above for the even-order moments. Hence the order of magnitude of any $(2m + 1)$th moment is at most n^{3m+1}. The $2m$th moments were shown to be of order n^{3m}. A $(2m + 1)$th moment is of order at most $n^{3/2(2m+1)}n^{-1/2}$. The odd moments are therefore of lower order $(n^{-1/2})$ compared with the even moments.

Finally, it follows from (5.38) that the moment μ_{rs} is the coefficient of $t_1^r t_2^s$ in

$$\exp\frac{2}{n^3}(h_{11}t_1^2 + 2h_{12}t_1t_2 + h_{22}t_2^3). \tag{5.39}$$

This is the moment-generating function of a bivariate normal distribution. The result is proved.

5.14　It is of some interest to consider the product-moment correlation between r_s and t. In exactly the same manner as in the derivation of (5.10) we find, for the mean value $E(cc')$

$$E(cc') = \frac{4}{n(n-1)(n-2)}[\sum a_{ij}a'_{ik} - \sum a_{ij}a'_{ij}][\sum b_{ij}b'_{ik} - \sum b_{ij}b'_{ij}]$$

$$+ \frac{2}{n(n-1)}\sum a_{ij}a'_{ij}\sum b_{ij}b'_{ij}. \tag{5.40}$$

In the particular case when c relates to S and c' to $\sum d^2$ we find

$$\sum_{i,j,k} a_{ij}a'_{ik} = \tfrac{1}{6}n^2(n^2 - 1) \tag{5.41}$$

$$\sum_{ij} a_{ij}a'_{ij} = \tfrac{1}{3}n(n^2 - 1) \tag{5.42}$$

and on substitution in (5.40), we find

$$E(cc') = \tfrac{1}{9}n^2(n - 1)(n + 1)^2. \tag{5.43}$$

Thus the product-moment correlation between S and $\sum d^2$, which is the same as that between t and r_s, is given by

$$\frac{E(cc')}{\sqrt{[E(c^2)E(c'^2)]}} = \frac{2(n + 1)}{\sqrt{[2n(2n + 5)]}}. \tag{5.44}$$

For large n this tends to one. Even for moderate n it is quite close. For $n = 5$ it is 0.980 and for $n = 20$ it is 0.990.

Corrections for continuity

5.15   We turn now to the question of corrections for continuity, the rules for which were given in Sections 4.12 and 4.15.

(a)   Consider first the case where one ranking is untied and the other contains ties, and may in the extreme case be a dichotomy. We may imagine the untied ranking in the natural order and the other in any arbitrary order. If we interchange a pair of neighbouring members in the untied ranking the only scores affected are those involving both members. Either the two ranks in the second ranking are untied or they are tied. In the first case the score S changes by two units, in the latter it remains unchanged. However many ties there are in the second ranking (short of the whole set being completely tied) there must be one interchange of neighbouring members in the first ranking which changes S by 2. Thus all intervals between successive values of S in the distribution of S are two units, and the appropriate correction for continuity is one unit.

(b)   If now the first ranking consists entirely of ties of extent u and the second is dichotomised, a change of two neighbours from different tied groups can at the most—and will at least for some rankings of the second variate—change S by $2u$, and the continuity correction is u.

(c)   If both variates are dichotomised then, as in Section 3.14, $S = ad - bc$. The least change that can result is an increase or decrease of a unit in a, and in such a case (say an increase) the increase in S is

$$(a + 1)(d + 1) - (b - 1)(c - 1) - (ad - bc) = a + b + c + d = n.$$

The continuity correction is thus $\frac{1}{2}n$.

(d)   When both rankings contain ties it is not possible to lay down any general rule for continuity corrections. If the point is important some special consideration such as that in Example 4.5 is necessary.

5.16   For the continuity corrections to r_s, consider a ranking in the natural order and a dichotomy into k and $n - k$ individuals, the corresponding midranks being $\frac{1}{2}(k + 1)$ and $\frac{1}{2}(n + k + 1)$. If two members of the first ranking are x and y and their interchange results in the interchange of two members in the second, one from each part of the dichotomy, the change in $\sum d^2$ is

$$[x - \tfrac{1}{2}(k + 1)]^2 + [y - \tfrac{1}{2}(n + k + 1)]^2 - [x - \tfrac{1}{2}(n + k + 1)]^2$$
$$- [y - \tfrac{1}{2}(k + 1)]^2 = n(x - y).$$

Thus there will be one interchange of neighbouring members ($y = x + 1$) which increases or decreases $\sum d^2$ by n. The appropriate deduction is therefore $\frac{1}{2}n$.

If the members of the first ranking are all tied to extent u, the minimum (and realisable) change is nu, giving a correction of $\frac{1}{2}nu$.

Finally, if the first ranking is dichotomised the minimum change is $\frac{1}{2}n^2$ and the correction is $\frac{1}{4}n^2$.

The non-null case

5.17   We turn now to the more difficult case where parental correlation τ is not zero. We prove first that the mean value of t over all possible samples is τ.

Consider the $\binom{N}{n}$ samples of n from a population of N members. Any particular pair of members will occur in $\binom{N-2}{n-2}$ samples, that is, all pairs occur equally frequently in the totality of all samples. Thus the total score for all samples is $\binom{N-2}{n-2}$ times the score for the population, say Σ. Thus

$$E(t) = \frac{\binom{N-2}{n-2}\Sigma}{\frac{1}{2}n(n-1)\binom{N}{n}} = \frac{\Sigma}{\frac{1}{2}N(N-1)} = \tau. \tag{5.45}$$

5.18   Next we derive an expression for the variance of t. Let $c^{(n)}$ be the quantity c for a sample ranking of n, and c be the parent value. Then

$$t = \frac{c^{(n)}}{n(n-1)} \tag{5.46}$$

and

$$c^{(n)} = \Sigma^{(n)} c_{ij}, \tag{5.47}$$

where $\Sigma^{(n)}$ denotes summation over those values of i and j occurring in the sample.

We require $E(t^2)$, so consider

$$\sum_n \{c^{(n)}\}^2 = \sum_n \Sigma\Sigma^{(n)} c_{ij}c_{kl}, \tag{5.48}$$

where Σ denotes summation over all selections of the sample of n from the population of N members. Let us enumerate the number of ways in which $c_{ij}c_{kl}$ and similar products with tied suffixes occur in the sum.

(i) When i, j, k, l are all different, the term $c_{ij}c_{kl}$ may occur with $\binom{N-4}{n-4}$ selections of the remaining members of the sample, and the contribution of such terms to Σ is

$$\binom{N-4}{n-4} \Sigma' c_{ij}c_{kl},$$

Σ' as usual denoting summation over unequal values of i, j, k, l from 1 to N.

(ii) The term $c_{ij}c_{il}$ similarly occurs in $\binom{N-3}{n-3}$ ways and there are four ways of tying one suffix. The contribution is thus

$$4\binom{N-3}{n-3} \Sigma' c_{ij}c_{il}.$$

(iii) Terms like c_{ij}^2 similarly contribute

$$2\binom{N-2}{n-2} \Sigma' c_{ij}^2.$$

Thus

$$\sum_n \{c^{(n)}\}^2 = \binom{N-4}{n-4} \Sigma' c_{ij} c_{kl} + 4\binom{N-3}{n-3} \Sigma' c_{ij} c_{ik}$$

$$+ 2\binom{N-2}{n-2} \Sigma' c_{ij}^2.$$

Expressing the Σ's in terms of Σs and dividing by $\binom{N}{n}$ we find

$$E\{c^{(n)}\}^2 = \frac{n^{[4]}}{N^{[4]}} (\Sigma c_{ij} c_{kl} - 4 \Sigma c_{ij} c_{il} + 2 \Sigma c_{ij}^2)$$

$$+ \frac{4n^{[3]}}{N^{[3]}} (\Sigma c_{ij} c_{il} - \Sigma c_{ij}^2) + \frac{2n^{[2]}}{N^{[2]}} \Sigma c_{ij}^2, \qquad (5.49)$$

where $n^{[r]} = n(n - 1) \cdots (n - r + 1)$. Since $\Sigma c_{ij}^2 = N(N - 1)$ and $\Sigma c_{ij} c_{kl} = c^2$ the variance of t for given τ and n is seen to depend on $\Sigma c_{ij} c_{il} = \Sigma c_i^2$, where $c_i = \Sigma_{j=1}^N c_{ij}$.

Let N become large. The quantities c and Σc_i^2 are respectively of order N^2 and N^3, so if we write

$$\tau_i = \frac{c_i}{N}, \qquad (5.50)$$

we find

$$E(t^2) \sim \frac{(n - 2)(n - 3)}{n(n - 1)} \tau^2 + \frac{4(n - 2)}{n(n - 1)} \frac{\Sigma \tau_i^2}{N} + \frac{2}{n(n - 1)}. \qquad (5.51)$$

Thus, in the limit,

$$\text{var } t = \frac{4(n - 2)}{n(n - 1)} \text{var } \tau_i + \frac{2}{n(n - 1)} (1 - \tau^2). \qquad (5.52)$$

5.19 Consider now $c_{ij} = a_{ij} b_{ij}$. Keeping b_{ij} equal to ± 1, let the as assume any values, subject to the conditions

$$\Sigma a_{ij}^2 = N(N - 1), \qquad \Sigma a_{ij} b_{ij} = c = N(N - 1)\tau.$$

The stationary values of Σc_i^2 occur when the as satisfy the relations

$$b_{ij}(c_i + c_j) - \lambda a_{ij} - \mu b_{ij} = 0 \qquad (5.53)$$

where λ and μ are undetermined multipliers. Multiplying by b_{ij} and summing for all j, we find

$$c_i = \frac{\mu(N - 1) - c}{N - 2 - \lambda}.$$

Thus, unless the c_i are all to be equal, in which case $\sum c_i^2$ is a *minimum*, λ and μ must take the values

$$\lambda = N - 2, \qquad \mu = c/(N - 1).$$

Multiplying (5.53) by a_{ij} and summing over j and i, we have

$$2 \sum c_i^2 - \lambda N(N - 1) - \mu c = 0,$$

whence it follows that $\sum c_i^2$ cannot exceed

$$\tfrac{1}{2}N(N - 1)(N - 2) + \tfrac{1}{2}c^2/(N - 1).$$

For large N this implies that

$$\sum \tau_i^2/N \le \tfrac{1}{2}(1 + \tau^2).$$

Hence

$$\text{var } \tau_i \le \tfrac{1}{2}(1 - \tau^2)$$

and thus, from (5.52),

$$\text{var } t \le \left\{ \frac{2(n - 2)}{n(n - 1)} + \frac{2}{n(n - 1)} \right\}(1 - \tau^2)$$

$$\le \frac{2}{n}(1 - \tau^2), \tag{5.54}$$

which is equation (4.12) of the previous chapter.

5.20   The form of this result suggests using a transformation

$$w = \sin^{-1} t. \tag{5.55}$$

To the same order of approximation we may take w as being normally distributed about $\omega = \sin^{-1} \tau$, and the variance of w will obey the relation

$$\text{var } w \le \frac{2}{n} \tag{5.56}$$

which has the advantage of being independent of ω.

5.21   The proof that the distribution of t tends to normality for large n follows, in essentials, the demonstrations given earlier in this chapter. We will merely outline it.

Write

$$g_{ij} = c_{ij} - c/N^2$$

so that

$$g_{ij} = g_{ji}, \qquad \sum g_{ij} = 0 \quad \text{and} \quad g_{ii} = -c/N^2 = -(N - 1)\tau/N.$$

The rth moment of $c^{(n)}$ about its mean value is $E\{\sum^{(n)} g_{ij}\}^r$, so consider

$$\sum_n \{\textstyle\sum^{(n)} g_{ij}\}^r = \sum_n \sum^{(n)} g_{ij} g_{kl} g_{uv} \cdots. \tag{5.57}$$

An essential condition is that

$$N^{-3} \sum g_{ij} g_{ik} \text{ has a non-zero bound,} \qquad (5.58)$$

which is true only if $1 - \tau^2$ is of the order of 1, so that the tendency to normality may break down for high correlations. We also assume that n/N tends to zero.

In the manner of Section 5.13, for the moment of order $2m$ the major term arises from expressions like $(\sum g_{ij} g_{ik})^m$ and other terms are of lower order in m. With $3m$ suffixes assigned there are $\binom{N-3m}{n-3m}$ ways of selecting the remaining $n - 3m$ members, and the suffixes can be tied in

$$\frac{(2m)! \, 2^{2m}}{m! \, (2!)^m}$$

ways to give the same result. Dividing by $\binom{N}{n}$ and noting that for large N and n

$$\binom{N-3m}{n-3m} \bigg/ \binom{N}{n} \sim \frac{n^{3m}}{N^{3m}}$$

we find for the major term in the moment of order $2m$

$$\frac{n^{3m}}{N^{3m}} \frac{(2m)!}{m!} 2^m (\sum g_{ij} g_{ik})^m$$

which is of order n^{3m}. By the same argument terms with $f < 3m$ different suffixes are of order n^f and may be neglected. Thus,

$$\mu_{2m} \sim \frac{n^{3m}}{N^{3m}} \frac{(2m)!}{m!} 2^m (\sum g_{ij} g_{ik})^m$$

$$\sim \frac{(2m)!}{2^m m!} (\mu_2)^m. \qquad (5.59)$$

Further, μ_{2m+1} is of order $n^{-1/2}$ in comparison. The tendency to normality follows. The variance of $c^{(n)}$ is

$$\frac{4n^3}{N^3} (\sum g_{ij} g_{ik}) = 4n^3 \operatorname{var} \tau_i, \qquad (5.60)$$

and the variance of t accordingly is $(4/n) \operatorname{var} \tau_i$ which agrees with (5.52) to the order considered.

5.22 We now give reasons for supposing that the limits to the variance of t given by (5.54) cannot be narrowed very substantially.

Consider a ranking such as

$$5 \quad 2 \quad 3 \quad 1 \quad 6 \quad 7 \quad 8 \quad 9 \quad 4$$

The number of positive pairs is 26, so $t = 0.44$. Let us transform this so as to bring the 1 to the beginning of the ranking, but move the 9 so as to preserve the score at 26. The 1 passes over three members to go to the beginning and hence adds 3 to the score. The 9 must therefore proceed to the left over three members so as to subtract 3, and we reach the ranking

$$1 \quad 5 \quad 2 \quad 3 \quad 9 \quad 6 \quad 7 \quad 8 \quad 4$$

Proceeding similarly with the 2 we reach

$$1 \quad 2 \quad 5 \quad 9 \quad 3 \quad 6 \quad 7 \quad 8 \quad 4$$

Had the 9 been contiguous to the 1 and incapable of proceeding further to the left we should have moved the 8, and so on. Continuing the process we ultimately arrive at

$$1 \quad 2 \quad 3 \quad 4 \quad 9 \quad 8 \quad 7 \quad 6 \quad 5$$

All the lower numbers are in the right order and the others in the inverse order. We may call this the 'canonical order' for given S. It is not always possible to reduce a given ranking to canonical order, but there cannot be more than one individual out of place.

If the parent ranking is inverted τ becomes $-\tau$. We may reduce this to the canonical form and re-invert the result, so that the coefficient is again τ. This ranking we may call the 'inverse canonical form'.

5.23   Now consider the canonical case when there are N members together, R at the beginning in the right order and $N - R$ in the inverse order. If we select $n - j$ members from the R and j from the $N - R$ the value of S for the sample of n is $\frac{1}{2}n(n - 1) + \frac{1}{2}j(j - 1)$ and the relative frequency of $Q = \frac{1}{2}n(n - 1) - S$ is

$$\binom{R}{n - j}\binom{N - R}{j} \Big/ \binom{N}{n}.$$

Now suppose that N tends to infinity and R/N tends to a limit p. The relative frequency of Q $(=\frac{1}{2}j(j - 1))$ then tends to $\binom{n}{j}p^{n-j}q^j$ where $q = 1 - p$. The mean value of Q is then

$$\sum_0^n \frac{1}{2}j(j - 1)\binom{n}{j}p^{n-j}q^j = \frac{1}{2}n(n - 1)q^2$$

and since

$$t = 1 - \frac{2Q}{\frac{1}{2}n(n - 1)}$$

we must have

$$q = \{\tfrac{1}{2}(1 - \tau)\}^{1/2}.$$

The variance of Q is found to be

$$\text{var } Q = n(n - 1)pq^2[nq + \tfrac{1}{2}(1 - 3q)]$$

and so

$$\text{var } t = \frac{16pq^2[nq + \tfrac{1}{2}(1 - 3q)]}{n(n - 1)}. \tag{5.61}$$

If the inverted parent is reduced to canonical form, giving ratios p' and q', we shall have

$$q' = [\tfrac{1}{2}(1 + \tau)]^{1/2}$$

and

$$\text{var } t' = \frac{16p'q'^2\{nq' + \tfrac{1}{2}(1 - 3q')\}}{n(n - 1)}. \tag{5.62}$$

Then, since $q^2 + q'^2 = 1$,

$$\text{var } t' - \text{var } t = \frac{16(n - 2)}{n(n - 1)}(q' - q)(1 - q)(1 - q').$$

When τ is positive $q' > q$ and then var $t' >$ var t. Taking the inverse canonical ranking when $\tau > 0$ and the direct canonical ranking when $\tau < 0$, we find for the variance of t when n is large

$$\text{var } t \sim \frac{4\sqrt{2}}{n}(1 + |\tau|)^{3/2}[1 - \sqrt{\tfrac{1}{2}(1 + |\tau|)}]. \tag{5.63}$$

The ratio of this quantity to the upper limit $(2/n)(1 - \tau^2)$ varies from $2(\sqrt{2} - 1) = 0.83$ when $\tau = 0$ to 1 when $\tau = 1$. Evidently the upper limit to the variance cannot be much improved, since an actual parent ranking has been found whose variance approximates to it for all values of τ when n is not too small.

More exact treatment in the non-null case

5.24 In proving that t tends to normality in the non-null case we have neglected terms of order $1/\sqrt{n}$ and this suggests that the normal approxima-tion may hold good for large n but may be indifferent for small or moderate n. We will examine briefly the possibility of improving the approximation in the case of moderate n.

Looking again at (5.52),

$$\text{var } t = \frac{4(n - 2)}{n(n - 1)} \text{var } \tau_i + \frac{2}{n(n - 1)}(1 - \tau^2), \tag{5.64}$$

we see that the actual variance of t for large N depends on the unknown functions τ_i and τ. In the absence of exact knowledge of these quantities we may estimate them from the sample, using the sample values of c_i and c instead of the unknown parent values. In doing so, however, it is better to modify our formulae slightly so as to remove bias. The reader will recall that in the ordinary theory of statistics it is better to use the estimator $\Sigma (x - x)^2/(n - 1)$ for the variance, rather than the actual sample variance $\Sigma (x - \bar{x})^2/n$, because the average of the former over all samples is the parent variance. For similar reasons it is better not to substitute the sample values of c_i and c in (5.64) but to use a formula which, averaged over all samples, gives the exact form of var t. Such an unbiased formula is

$$\text{var } t = \frac{1}{n(n - 1)(n - 2)(n - 3)} \left[4 \Sigma c_i^2 - \frac{2(2n - 3)}{n(n - 1)} c^2 - 2n(n - 1) \right]$$

(5.65)

where the c_i and c values are sample values. This will give us a 'best' estimate of var t.

5.25   We may take matters a little further by considering the third moment of t so as to allow for departures from normality in the sampling distribution. Reference may be made to Daniels and Kendall (1947) for the details. We merely quote the result that if

$$\gamma_1 = \frac{\mu_3(t)}{[\mu_2(t)]^{3/2}},$$

(5.66)

then the frequency-distribution of

$$x = \frac{t - \tau}{\sqrt{\mu_2(t)}}$$

is

$$f(x) = \left(1 - \frac{\gamma_1}{6} \frac{d^3}{dx^3} \right) \frac{e^{-1/2x^2}}{\sqrt{(2\pi)}} [1 + 0(n^{-1})].$$

(5.67)*

If ξ is the normal deviate whose chance of being exceeded is $P(\xi)$, the chance of x exceeding ξ is

$$F(\xi) = P(\xi) + \frac{\gamma_1}{6} (\xi^2 - 1) \frac{e^{-1/2\xi^2}}{\sqrt{(2\pi)}}.$$

If X is the correct limit such that $F(X) = P(\xi)$ it is readily proved by successive approximation that

$$X = \xi + \frac{\gamma_1}{6} (\xi^2 - 1)$$

(5.68)

to order n^{-1}. For example the 5 per cent value of ξ is ± 1.96. The corresponding value of X is

$$\pm 1.96 + \frac{(1.96)^2 - 1}{6} \gamma_1 = \pm 1.96 + 0.474\gamma_1.$$

(5.69)

---

*This is a stronger result than would be obtained by the usual expansion of a frequency function in a Gram–Charlier series based on the first three moments only.

The corresponding 1 per cent value of X is

$$\pm 2.58 + 0.941 y_1. \tag{5.70}$$

The following example illustrates the use of these results.

Example 5.1

Table 5.1 shows the ranks according to two qualities X and Y in a sample of 30 drawn from a population of unknown character. We want to estimate the correlation in the parent population.

Table 5.1

X	Y	X	Y	X	Y
1	5	11	17	21	21
2	4	12	13	22	29
3	9	13	24	23	28
4	3	14	14	24	19
5	6	15	1	25	23
6	2	16	12	26	20
7	15	17	10	27	7
8	18	18	30	28	26
9	8	19	22	29	27
10	11	20	16	30	25

The correlation t is found to be $+0.490$.

(a)   Consider first of all the maximum confidence limits given by (4.15). The 5 per cent limits are

$$-0.02 \le t \le 0.80.$$

(b)   The $\sin^{-1}$ transformation of Section 5.20 gives

$$w = \sin^{-1} t = 0.512$$

$$0.01 \le t \le 0.85,$$

results which are not very different from those of (a).

(c)   To go further we require the values c_i and c. Table 5.2 shows the matrix of values c_{ij} for the data, and we find

$$c = 426$$

$$\sum c_i^2 = 7470.$$

From (5.65) we then have

$$\text{var } t = \frac{1}{30(29)(28)(27)} \left[4(7470) - \frac{2(57)}{30(29)} (426)^2 - 60(29) \right]$$

$$= 0.006\,630$$

Table 5.2

c_{ij}																														c_i
○	−	+	−	+	−	+	+	+	+	+	+	+	+	+	−	+	+	+	+	+	+	+	+	+	+	+	+	+	+	21
−	○	+	−	+	−	+	+	+	+	+	+	+	+	+	−	+	+	+	+	+	+	+	+	+	+	+	+	+	+	21
+	−	○	−	+	−	+	+	+	+	+	+	+	+	+	−	+	+	+	+	+	+	+	+	+	+	+	+	+	+	17
+	+	○	−	+	+	+	+	+	+	+	+	+	+	+	−	+	+	+	+	+	+	+	+	+	+	+	−	+	+	19
+	+	−	+	○	−	+	+	+	+	+	+	+	+	+	−	+	+	+	+	+	+	+	+	+	+	+	+	+	+	23
+	+	−	−	−	○	+	+	+	+	+	+	+	+	+	−	+	+	+	+	+	+	+	+	+	+	+	+	+	+	17
+	+	+	+	+	+	○	−	+	+	+	−	+	−	−	−	+	+	+	+	+	+	+	+	+	−	+	+	+	+	13
+	+	+	+	+	+	+	○	−	+	−	+	−	+	−	−	−	−	+	+	+	+	+	+	+	−	+	+	+	+	9
+	+	+	+	+	+	−	+	○	−	+	−	+	+	+	−	+	+	+	+	+	+	+	+	+	−	+	+	+	+	19
+	+	+	+	+	+	−	−	+	○	−	+	+	+	+	−	+	+	+	+	+	+	+	+	+	−	+	+	+	+	19
+	+	+	+	+	+	−	+	−	+	○	−	+	−	−	−	+	+	+	+	+	+	+	+	+	−	+	+	+	+	13
+	+	+	+	+	+	+	−	+	−	+	○	+	+	+	−	+	+	+	+	+	+	+	+	+	−	+	+	+	+	15
+	+	+	+	+	+	−	+	−	−	−	−	○	−	−	−	+	+	+	+	+	+	+	+	+	−	+	+	+	+	7
+	+	+	+	+	+	+	−	−	−	+	−	+	○	−	−	+	+	+	+	+	+	+	+	+	−	+	+	+	+	13
−	−	−	−	−	−	−	−	−	−	−	−	−	−	○	+	+	+	+	+	+	+	+	+	+	+	+	+	+	+	1
+	+	+	+	+	+	+	+	+	+	+	+	+	+	−	○	+	+	+	+	+	+	+	+	+	−	+	+	+	+	13
+	+	+	+	+	+	−	−	−	−	+	−	−	−	−	+	○	+	+	+	+	+	+	+	+	−	+	+	+	+	11
+	+	+	+	+	+	−	+	−	+	−	+	−	−	−	−	+	○	+	+	+	+	+	+	+	−	+	+	+	+	5
+	+	+	+	+	+	−	+	−	+	−	+	−	−	−	−	+	+	○	+	+	+	+	+	+	−	+	+	+	+	15
+	+	+	+	+	+	−	−	−	−	+	−	−	−	−	+	+	+	+	○	+	+	+	+	+	−	+	+	+	+	17
+	+	+	+	+	+	−	−	−	−	+	−	−	−	−	+	+	+	+	+	○	+	+	+	+	−	+	+	+	+	19
+	+	+	+	+	+	−	−	−	−	+	−	−	−	−	+	+	+	+	+	+	○	+	+	+	−	+	+	+	+	11
+	+	+	+	+	+	−	−	−	−	+	−	−	−	−	+	+	+	+	+	+	+	○	+	+	−	+	+	+	+	11
+	+	+	+	+	+	−	−	−	−	+	−	−	−	−	+	+	+	+	+	+	+	+	○	+	−	+	+	+	+	15
+	+	+	+	+	+	−	−	−	−	+	−	−	−	−	+	+	+	+	+	+	+	+	+	○	−	+	+	+	+	17
+	+	+	+	+	+	+	+	+	+	+	+	+	+	+	+	+	+	+	+	+	+	+	+	+	○	+	+	+	+	15
−	−	−	−	−	−	−	−	−	−	−	−	−	−	−	−	−	−	−	−	−	−	−	−	−	−	○	+	+	+	−11
+	+	+	+	+	+	+	+	+	+	+	+	+	+	+	+	+	+	+	+	+	+	+	+	+	+	−	○	+	−	21
+	+	+	+	+	+	+	+	+	+	+	+	+	+	+	+	+	+	+	+	+	+	+	+	+	+	+	−	○	−	21
+	+	+	+	+	+	+	+	+	+	+	+	+	+	+	+	+	+	+	+	+	+	+	+	+	+	−	+	+	○	19

$c = 426$
$n = 30$
$\Sigma c^2 = 7470$

giving an estimated standard error of 0.0814. The 5 per cent confidence limits, assuming normality, are then

$$0.33 \le t \le 0.65.$$

These are much narrower than the values of (a) and (b).

(d)   To allow for departures from normality we need to estimate γ_1, which in turn depends on $\mu_3(t)$. For moderate samples the following formula gives an approximation

$$\mu_3(t) = \frac{8}{n^6} \left[\sum_{i>j} c_{ij}(c_i + c_j)^2 - \frac{5c \sum c_i}{n} + \frac{3c^2}{n^3} \right], \tag{5.71}$$

where the first term in square brackets is a summation of $c_{ij}(c_i + c_j)^2$ over all values $i > j$, i.e. the values in Table 5.2 below the diagonal. After some tedious computation we find

$$\gamma_1 = -0.32.$$

The adjusted 5 per cent limits from (5.69) are then

$$0.32 \le t \le 0.64.$$

The corrections for non-normality are small, and the limits are very similar to those given by (c).

5.26   The arithmetic required by the foregoing examples is more than we should usually have the patience to perform. Sundrum (1953*b*) has investigated the third and fourth moments of *t* in the non-null case and has shown that the latter depends on 10 parameters which can be estimated from the data. This result is of considerable theoretical interest, but again the labour of applying it in practice would be prohibitive unless it could be programmed on a high-speed computer.

One further resource is left to us. If we can assume that the rankings are based on a normal variate (and, presumably, as an approximation if the underlying variate is nearly normal) the variance is narrowed very considerably. See Sections 9.6–9.19.

r_s in the non-null case

5.27   As usual, the sampling theory of r_s is more complicated than that of *t* in the non-null case; here we derive the expected value of r_s.

Let us consider the function *V* of (2.27),

$$V = \sum_{i<j} m_{ij}(j - i). \tag{5.72}$$

The value of m_{ij} for any particular pair of members is the same for both sample and population—this is the basic reason why $E(t) = \tau$. But the factor $(j - i)$ is not.

Consider a pair of members having ranks I, J in the population of N. Any member between them in the population has probability $(n - 2)/(N - 2)$ of appearing in the sample ranking. The mean number of such ranks appearing in the sample is therefore

$$\frac{n - 2}{N - 2}(J - I - 1)$$

and hence the mean $j - i$ in the sample is one more than this, namely

$$\frac{N - n}{N - 2} + \frac{n - 2}{N - 2}(J - I). \tag{5.73}$$

Thus the mean value of V is given by

$$E(V) = \frac{\binom{N - 2}{n - 2}}{\binom{N}{n}} \sum_{I < J} m_{IJ}\left[\frac{N - n}{N - 2} + \frac{n - 2}{N - 2}(J - I)\right],$$

where summation takes place over the population; for any pair of ranks can occur in $\binom{N - 2}{n - 2}$ ways. Hence, expressing V in terms of r_s we find

$$E(1 - r_s) = \frac{12}{N(N - 1)(n + 1)} \sum m_{IJ}\left[\frac{N - n}{N - 2} + \frac{n - 2}{N - 2}(I - J)\right]. \tag{5.74}$$

Now for our population

$$\sum m_{IJ} = \frac{N(N - 1)}{4}(1 - \tau)$$

$$\sum m_{IJ}(J - I) = \frac{N(N^2 - 1)}{12}(1 - \rho_s)$$

and substituting in (5.74) we find

$$E(r_s) = \frac{1}{(n + 1)(N - 2)}[3(N - n)\tau + (n - 2)(N + 1)\rho_s], \tag{5.75}$$

a general formula obtained independently by Durbin and Stuart (1951) and Daniels (1951).

5.28 We note that for N large this tends to

$$E(r_s) = \frac{1}{n + 1}[3\tau + (n - 2)\rho_s] \tag{5.76}$$

a result obtained for continuous populations by Hoeffding in 1948. It follows that

$$E(r_s) - \rho_s = \frac{3}{n + 1}(\tau - \rho_s) \tag{5.77}$$

and r_s is a biased estimator of ρ_s and for some populations the bias may be appreciable. It is probably well to correct r_s by deducting the 'sample bias' $3(t - r_s)/(n - 2)$.

Notes and references

See the notes and references to Chapter 4.

For the mean value of rho see Hoeffding (1948a), Daniels (1950, 1951), Durbin and Stuart (1951). For a sample from a bivariate normal distribution with non-zero correlation, the mean was found by Moran (1948b) and the variance for large n by Kendall (1949) and David, Kendall and Stuart (1951). More exact values for the variance were found by David and Mallows (1961). The distribution of rho was investigated by Fieller, Hartley and Pearson (1957), Fieller and Pearson (1961), Pearson and Snow (1962). More recently, Kraemer (1974) found a transformation that gives more accurate approximations than the Fisher z-transformation; this transformation simplifies calculation of critical values and power. The asymptotic normality of rho was established under all distributions by Govindarajulu (1976).

A simpler proof of the result in Section 5.8 is given in Moran (1950b) and Silverstone (1950). These proofs depend on more sophisticated ideas and do not generalise to other rank coefficients, but they give the moments of the distribution directly.

The mean and variance of tau were found by Esscher (1924) and Greiner (1909). Kerridge (1975) gave a probability interpretation of the mean of tau. Improved bounds on the variance of tau were found by Stuart (1956a). Sundrum (1953d) gave a proof of the result that the canonical ranking discussed in Section 5.22 has a minimum variance.

Hoeffding (1947) discussed the distribution of tau in the non-null case. The normality of tau in the presence of ties follows as a consequence of general results obtained by Hoeffding (1948a). A simple proof is not easy to give, but see Terpstra (1952) for the case when only one ranking is tied or Kruskal (1952). The asymptotic normality of tau was studied by Best (1974) and Jirina (1976).

Noether (1967) gave an extensive treatment of the theoretical properties of tau, including a confidence interval estimate. Snow (1962, 1963) investigated the distribution of tau for samples from a bivariate normal distribution.

Wood (1970) gave transformations that stabilise the variances of both rho and tau. Goodman and Kruskal (1972) gave a unified method of deriving variances for tau, rho and other measures of association. Ruymgaart (1973) developed further asymptotic theory.

Goodman and Kruskal (1954, 1959, 1963, 1972) explored many properties of rho and tau. The 1963 and 1972 papers discussed asymptotic theory and the adequacy of normal approximations. These papers are collected in Goodman and Kruskal (1979).

The power and efficiency of both rho and tau were studied by Stuart (1954b, 1954c), Konijn (1956), Farlie (1960, 1961, 1963), Gokhale (1968),

Mardia (1969), Bhattacharyya, Johnson and Neave (1970), Woodworth (1970), Moran (1979), among others. Their asymptotic efficiency relative to the test based on Pearson's r is $9/\pi^2$ for the bivariate normal distribution. Markowski (1987) compares rho and tau on the basis of their efficacies.

Schweizer and Wolff (1981) give eight desirable properties for nonparametric measures of dependence and compare and contrast rho, tau and other coefficients with respect to these axioms.

Hoeffding (1984b) presented a test of independence that is sensitive to broad classes of dependency. Blum, Kiefer and Rosenblatt (1961) gave a related test that is asymptotically equivalent. Bhuchongkul (1964) and Bell and Doksum (1967) both studied general classes of tests for independence. Schemper (1984b) gave an exact test for generalised Kendall correlation coefficients.

The non-null distribution of Spearman's rho is very complicated. Kraemer (1974) and Henze (1979) obtained some useful results.

Chapter 6

The problem of *m* rankings

6.1 Up to this point we have been concerned with the correlation between two rankings. We now consider the case when there are several rankings, say *m* in number, of *n* objects and we desire to investigate the general relationship between the rankings. Suppose, for instance, four observers rank six objects as shown in Table 6.1.

Table 6.1

	Object					
	A	B	C	D	E	F
Observer *W*	5	4	1	6	3	2
Observer *X*	2	3	1	5	6	4
Observer *Y*	4	1	6	3	2	5
Observer *Z*	4	3	2	5	1	6
Totals of ranks:	15	11	10	19	12	17

In accordance with our known methods we can work out the rank correlation coefficient between each pair of observers, obtaining $\binom{4}{2} = 6$ coefficients. This, however, is not what we usually require. We need a measure of the concordance of the four observers taken as a group.

The most obvious procedure would be to average the six coefficients between pairs of observers but this would be tedious for a larger number of observers. We will develop instead a measure of the concordance of the four observers taken as a group.

Our measure will be based on the sum of squares of deviations of the column totals in Table 6.1 around their mean, which is $84/6 = 14$. We call this sum S and obtain

$$S = (15 - 14)^2 + (11 - 14)^2 + \cdots + (17 - 14)^2 = 64.$$

If the ranks had been allotted at random by each observer, the column totals in (6.1) would tend to be all equal to 14 and S would then equal zero. On the other hand, if a strict ordering of the objects existed and all observers agreed upon that ordering, the best object would have four ranks of 1, and so on.

Then the rank sums in Table 6.1 would be some permutation of 4, 8, 12, 16, 20, 24 and the corresponding sum of squares of deviations would be as large as possible as

$$\max S = (4 - 14)^2 + (8 - 14)^2 + \cdots + (24 - 14)^2 = 280.$$

We define the measure of relative agreement as

$$W = \frac{S}{\max S} = \frac{64}{280} = 0.229. \tag{6.1}$$

6.2 In general with n objects and m rankings or observers the total of all ranks is $mn(n + 1)/2$ and so the average column sum in a table like Table 6.1 is $m(n + 1)/2$. Let $R_1, R_2, \ldots, R_n$ denote the actual column rank sums. Then the sum of squares of deviations is

$$S = \sum_{i=1}^{n} \left[R_i - \frac{m(n + 1)}{2} \right]^2 = \sum_{i=1}^{n} R_i^2 - \frac{nm^2(n + 1)^2}{4}. \tag{6.2}$$

If all the rankings were in complete agreement the rank sums would be some permutation of

$$1m, 2m, \ldots, nm$$

and

$$\max S = \sum_{i=1}^{n} \left[im - \frac{m(n + 1)}{2} \right]^2 = \frac{m^2(n^3 - n)}{12}. \tag{6.3}$$

The measure of relative agreement then is

$$W = \frac{12S}{m^2(n^3 - n)}, \tag{6.4}$$

which we call the Kendall coefficient of concordance.

6.3 W measures, in a sense, the communality of judgements for the m observers. If they all agree $W = 1$. If they differ very much among themselves the sums of ranks will be more or less equal, and consequently the sum of squares S becomes small compared with the maximum possible value, so that W is small. As W increases from 0 to 1 the deviations become 'more different' and there is a greater measure of agreement in the rankings.

6.4 The reader may wonder why we have chosen a coefficient ranging from 0 to 1 and not from -1 to 1 as for a rank correlation coefficient. The answer is that when more than two observers are involved, agreement and disagreement are not symmetrical opposites. The m observers may all agree but they cannot all disagree completely, in the sense here considered. If, of three observers W, X and Y, W disagrees with X on a comparison and also disagrees with Y, then X and Y must agree.

6.5 If we write r_{av} for the mean value of the Spearman coefficients between the $\binom{m}{2}$ possible pairs of observers, then

$$r_{av} = \frac{mW - 1}{m - 1}. \tag{6.5}$$

For if the rank of the jth object by the ith observer, measured from the mean $\frac{1}{2}(n + 1)$, is x_{ij} the average of all the rank correlations is

$$
\begin{aligned}
r_{av} &= \frac{1}{m(m - 1)} \frac{\displaystyle\sum_{i,k=1}^{m} \sum_{j=1}^{n} x_{ij} x_{kj}}{\frac{1}{12}(n^3 - n)}, \qquad i \neq k \\[2mm]
&= \frac{12}{m(m - 1)(n^3 - n)} \left[\sum_{j=1}^{n} \left(\sum_{i=1}^{m} x_{ij} \right)^2 - \sum_{j=1}^{n} \sum_{i=1}^{m} x_{ij}^2 \right] \\[2mm]
&= \frac{12}{m(m - 1)(n^3 - n)} \left[S - \frac{1}{12} m(n^3 - n) \right] \\[2mm]
&= \frac{mW - 1}{m - 1}. \tag{6.6}
\end{aligned}
$$

When $r_{av} = +1$, $W = 1$. When $W = 0$, $r_{av} = -1/(m - 1)$, and this is the least value which this average can take, a further illustration of the point made at the end of the last section.

6.6 If some of the rankings contain ties we may write, as in (3.6),

$$U' = \tfrac{1}{12} \sum (u^3 - u). \tag{6.7}$$

In this case we shall define the coefficient W as

$$W = \frac{S}{\frac{1}{12}m^2(n^3 - n) - m \sum U'}. \tag{6.8}$$

the summation $\sum$ taking place over all sets of tied ranks.

In this case (6.6) requires some modification.

6.7 This definition requires a little comment, for the denominator in (6.8) is not necessarily the maximum value of the sum of squares of deviations of totals of ranks (from their mean). We may, in fact, define W in the untied case by the alternative formula

$$W = \frac{S}{mS'}, \tag{6.9}$$

where S' is the sum of squares of deviations of all individual ranks from their mean. This is in agreement with our previous definition in (6.4), since the sum of squares of deviations in any ranking is $\frac{1}{12}(n^3 - n)$ and there are m rankings. The definition (6.9) is also in accordance with (6.8), for, as we have seen in Section 3.10, the effect of ties in a ranking providing a number U' is to reduce the sum of squares of deviations by U'.

The reason for adopting (6.9) for the definition is that it bears an analogy to the analysis of variance. Suppose we array the ranks (measured from their mean) as

$$
\begin{array}{cccc}
x_{11} & x_{12} & \cdots & x_{1n} \\
x_{21} & x_{22} & \cdots & x_{2n} \\
\vdots & \vdots & & \vdots \\
x_{m1} & x_{m2} & \cdots & x_{mn}.
\end{array}
\tag{6.10}
$$

The sums of rows are all zero with corresponding zero mean. The sums of columns may be written $S_1, \ldots, S_n$ with means $S_1/m \ldots, S_n/m$.

Then the variance of the whole array is S'/mn by definition. The variance of column means is $(1/n) \sum (S_i/m)^2 = S/m^2n$. The ratio of the two is thus $S/mS' = W$, which is thus exhibited as the ratio of the variance of column means to the whole variance. This, and still more the ratio $S/(mS' - S)$, is a familiar ratio in the analysis of variance.

Example 6.1

Consider the three rankings shown in Table 6.2.

Table 6.2

X	1	$4\frac{1}{2}$	2	$4\frac{1}{2}$	3	$7\frac{1}{2}$	6	9	$7\frac{1}{2}$	10
Y	$2\frac{1}{2}$	1	$2\frac{1}{2}$	$4\frac{1}{2}$	$4\frac{1}{2}$	8	9	$6\frac{1}{2}$	10	$6\frac{1}{2}$
Z	2	1	$4\frac{1}{2}$	$4\frac{1}{2}$	$4\frac{1}{2}$	$4\frac{1}{2}$	8	8	8	10
Total:	$5\frac{1}{2}$	$6\frac{1}{2}$	9	$13\frac{1}{2}$	12	20	23	$23\frac{1}{2}$	$25\frac{1}{2}$	$26\frac{1}{2}$

The mean is $16\frac{1}{2}$ and the deviations from this mean are

$$-11 \quad -10 \quad -7\frac{1}{2} \quad -3 \quad -4\frac{1}{2} \quad 3\frac{1}{2} \quad 6\frac{1}{2} \quad 7 \quad 9 \quad 10$$

The sum of squares of deviations is $S = 591$.

For the U'-numbers we have

$$X: \tfrac{1}{12}[2(2^3 - 2)] = 1$$

$$Y: \tfrac{1}{12}[3(2^3 - 2)] = 1\tfrac{1}{2}$$

$$Z: \tfrac{1}{12}(4^3 - 4 + 3^3 - 3) = -7.$$

Thus, from (6.8),

$$W = \frac{591}{742.5 - 28.5}$$

$$= 0.828.$$

The effect of taking ties into account is evidently small.

The significance of *W*

6.8 We now consider testing the significance of an observed value of *W*. If all the observers are independent in their judgments, then any set of rankings is just as probable as any other set. We shall therefore consider the distribution of *W* in the $(n!)^m$ possible sets of ranks and use it in the customary way to reject or accept the null hypothesis that the observers have no community of preference or no association between rankings.

The actual distribution of *S* and therefore *W* is given in Appendix Table 5 for $n = 3$, $m = 2$ to 10; $n = 4$, $m = 2$ to 6; $n = 5$, $m = 3$.

Example 6.2

Shearer (1982) reported a set of ratings made by newsmen representing CBS, NBC and ABC about the press conference behaviour of recent US presidents. The last seven presidents were rated on a scale of 1 to 10 (10 = best) on four different attributes as shown in Table 6.3.

Table 6.3

	Candour	Informative value	Combative skill	Humour
Reagan	6	4	8	8.5
Carter	8	7	6	7
Ford	7	5	4	6
Nixon	5	6	8	4
Johnson	2	3	6	4
Kennedy	5	6	9	9
Eisenhower	8	7	4	3

If we want to investigate the overall ratings of the presidents, we should rank these ratings from 1 to 7 for each attribute separately (1 = best) and find the totals as shown in Table 6.4.

We have $m = 4$ rankings of $n = 7$ objects and $\sum R_i^2 = 1921$ and $S = 129$. The correction in (6.7) is

$$\text{Candour: } 2(2^3 - 2)/12 = 1.0$$
$$\text{Informative: } 2(2^3 - 2)/12 = 1.0$$
$$\text{Combative: } 3(2^3 - 2)/12 = 1.5$$
$$\underline{\text{Humour: }\quad (2^3 - 2)/12 = 0.5}$$
$$4.0$$

From (6.8) we obtain $W = 0.2986$. From Appendix Table 5, $S = 129$ is not significant at the 0.05 level, so there is no significant agreement between the rankings.

For larger values of *m* and *n*, we have the following two approximations to the distribution of *S*.

Table 6.4

	Candour	Informative	Combative	Humour	Total
Reagan	4	6	2.5	2	14.5
Carter	1.5	1.5	4.5	3	10.5
Ford	3	5	6.5	4	18.5
Nixon	5.5	3.5	2.5	5.5	17.0
Johnson	7	7	4.5	5.5	24.0
Kennedy	5.5	3.5	1	1	11.0
Eisenhower	1.5	1.5	6.5	7	16.5

(1)   For all values other than those in Appendix Tables 5 an approximation may be based on Fisher's z-distribution with v_1 and v_2 degrees of freedom. We write

$$z = \tfrac{1}{2} \log_e \frac{(m-1)W}{1-W} \tag{6.11}$$

with degrees of freedom

$$v_1 = n - 1 - \frac{2}{m} \tag{6.12}$$

$$v_2 = (m-1)v_1.$$

Fisher's z-distribution is given here as Appendix Tables 7A and 7B for probability levels 0.05 and 0.01 respectively. Direct use of the tables may be obviated by using Appendix Table 6 which gives the corresponding values of S to those of z at probability levels of 5 per cent and 1 per cent for various values of m from 3 to 20 and for n from 3 to 7. The use of this table will be illustrated in a moment.

(2)   Although the above test is generally valid, a simpler test may be used for $n > 7$. If we write

$$\chi_r^2 = m(n-1)W = \frac{S}{\tfrac{1}{12}mn(n+1)} \tag{6.13}$$

then the distribution of χ_r^2 is approximately χ^2 with $v = n - 1$ degrees of freedom, given here as Appendix Table 8.

Example 6.3

For $m = 18$ rankings of $n = 7$ objects, we find $S = 1620$ so that

$$W = \frac{1620}{\tfrac{1}{12}(18^2)(336)} = 0.179.$$

Appendix Table 5 does not cover $m = 18$, $n = 7$, so we use the approximation based on Fisher's z-distribution. From Appendix Table 6 we find the critical values for the 5 per cent level as

$$m = 15 \qquad S = 864.9$$
$$m = 20 \qquad S = 1158.7$$

and for the 1 per cent level as

$$m = 15 \qquad S = 1129.5$$
$$m = 20 \qquad S = 1521.9.$$

For $m = 18$ the appropriate values of S lie between the values for $m = 15$, $m = 20$, and our observed value is greater than the value for 1 per cent. This means that the probability of obtaining a value as great as or greater than the observed value is less than 0.01—the value lies, we may say, beyond the 1 per cent point. It is thus significant if we agree that such small probabilities are significant.

Example 6.4

In 28 rankings of 13 a value of S was found of 11 440 and hence

$$W = \frac{11\,440}{\frac{1}{12}(28^2)(2184)} = 0.080.$$

We may test this by the use of the chi-square approximation in (6.13). We have

$$\chi_r^2 = \frac{11\,440}{\frac{1}{12}(28)(13)(14)} = 27.$$

From Appendix Table 8 we see that for $v = n - 1 = 12$ degrees of freedom, at the 1 per cent significance level, $\chi^2 = 26.217$. Our observed value is slightly greater than this and is thus 'just significant' at the 1 per cent point.

Continuity correction for *W*

Example 6.5

In practical cases the number m is often so large that no correction for continuity need be made, but if one is desirable, it may be introduced by subtracting unity from S and adding 2 to max S to obtain

$$W = \frac{S - 1}{\frac{1}{12}m^2(n^3 - n) + 2}.$$

For instance, consider the case $n = 3$, $m = 9$, and suppose $S = 78$. From Appendix Table 5A we see that the probability of such a value or greater is 0.010, so that this is approximately the 1 per cent point. Suppose we apply Fisher's z-test to these data. We find, with continuity corrections,

$$W = \frac{78 - 1}{\frac{1}{12}(81)(24) + 2} = 0.4695$$

$$z = 0.979, \qquad v_1 = \frac{16}{9}, \qquad v_2 = \frac{128}{9}.$$

By linear interpolation of reciprocals in Appendix Table 7B we find, for these values of v_1 and v_2, a value of z equal to 0.954 against the exact value of 0.979. Even for such low values of n as 3 the approximation given by the z-test is fair.

6.9   When ties are present we use average ranks. Fisher's z-test requires no modification unless the number or extent of the ties is large.

In the latter case the test becomes more complicated. Let μ_{2i} be the variance of the ith ranking typified by $\frac{1}{12}(n^2 - 1) - (1/n)U'$.

Write

$$\mu_2(W) = \frac{4}{m^2(n - 1)} \frac{\sum\limits_{i,j} \mu_{2i}\mu_{2j}}{(\sum \mu_{2i})^2}, \tag{6.14}$$

the summation extending over the $\frac{1}{2}m(m - 1)$ values $i \neq j$. Then W may be tested with z given by (6.11) and the modified degrees of freedom

$$v_1 = \frac{2(m - 1)}{m^3\mu_2(W)} - \frac{2}{m} \tag{6.15}$$

$$v_2 = (m - 1)v_1.$$

The appropriate value of χ_r^2 is

$$\chi_r^2 = \frac{S}{\frac{1}{12}mn(n + 1) - [1/(n - 1)] \sum U'}. \tag{6.16}$$

Estimation

6.10   Suppose now that a value of W has been found to be significant, so that there is evidence of some agreement among the observers. If we go further and suppose that their judgments are more or less accurate according to some objective scale we may ask: what is the true ranking of the objects, or rather, what is the best estimate we can make of that true ranking?

6.11   Suppose we have three rankings of eight as shown in Table 6.5.

Table 6.5

	Object							
	A	B	C	D	E	F	G	H
Observer X	4	2	1	7	6	3	5	8
Observer Y	7	2	1	6	4	5	3	8
Observer Z	7	4	2	6	5	3	1	8
Totals:	18	8	4	19	15	11	9	24

One procedure which we must dismiss is that of ranking according to the number of 'firsts', 'seconds', etc., obtained by each individual. For instance we might rank C first because it has two 'firsts'. Object G has the remaining 'first' and we might rank it second. Looking then to the 'seconds' we find that B has two so we rank it third. The other second occurred under C, which has already been ranked, so we proceed to the 'thirds', and so on. The ranking obtained in this way is

$$C \quad G \quad B \quad F \quad E \quad A \quad D \quad H. \tag{6.17}$$

When we consider the 'fourths', there are two members, E and A, having one each, but we give precedence to the former because it has a fifth whereas the latter has only two 'sevenths'.

This procedure is not self-consistent. Suppose we start from the other end of the ranking and rank as 8 the individual with the greatest number of 'eighths' and so on. Then in our present example we get

$$C \quad B \quad G \quad F \quad E \quad D \quad A \quad H, \tag{6.18}$$

which is not the same as before. In general there is no particular reason for starting at one end rather than at the other, and it is evidently unsatisfactory that the two procedures should give different results.

6.12 A better procedure is to rank according to the sums of ranks allotted to the individuals. Thus, in Table 6.5 C has the least total, so we rank it first, B has next lowest, and so on. The ranking thus obtained is

$$C \quad B \quad G \quad F \quad E \quad A \quad D \quad H, \tag{6.19}$$

which is different from either (6.17) or (6.18).

It may be shown (as in the next chapter) that this gives a 'best' estimate in a certain sense associated with least squares. In fact, the sum of squares of differences between what the totals are and what they would be if all rankings were alike is a minimum when the ranking is estimated by this method. Furthermore, if the ranking arrived at by this method is correlated by Spearman's r_s with the observed rankings, the mean of these rank correlations is larger than for any other estimated ranking. This is not necessarily true for t also but will usually be so for rankings of moderate size.

6.13 A few points require mention in connection with ties. Even if no observed ranking contains ties, the object rank totals or column sums may contain ties. This will produce ambiguities in the estimated ranking.

Suppose that ten objects are ranked by four observers and the rankings of some three particular objects A, B and C are as shown in Table 6.6. Since the totals of the ranks for objects A, B and C are all the same, our method of estimation gives no criterion of choice. If ties are permitted in the estimated ranking, the three objects should all be given the same ranking.

Table 6.6

	Object		
	A	B	C
Observer W	7	8	10
Observer X	9	8	6
Observer Y	3	7	6
Observer Z	5	1	2
Totals	24	24	24

If ties are not permitted, it seems best to give preference according to how closely the ranks given cluster around their mean rank, i.e., according to the variances of the ranks. Since the mean ranks are all equal when the column totals are equal, this procedure is equivalent to giving precedence according to the sums of squares of the ranks. In Table 6.6 the sums of squares for A, B, C are 164, 178, 176 respectively. The ranking according to this procedure is that A is best, followed by C, followed by B.

If an individual rank j is regarded as expressing the fact that $j - 1$ objects are preferred to it by that observer, the sum of ranks gives the sum of preferences when m, the number of rankings or observers, is subtracted. In the example above, objects A, B and C were each preferred to the other 9 objects $24 - 4 = 20$ times. Our method of estimation ranks the objects according to the number of preferences. If some ranks are tied, the replacement of integer ranks by average ranks does not affect the number of preferences to other objects and merely cancels preferences within the tied set. Thus our method continues to rank according to preferences even when ties exist.

Example 6.6

6.14 Lawrence (1983) designed an experiment to compare the usefulness of three popular forecasting techniques—judgmental forecasting, Box–Jenkins methods and exponential smoothing. Each of eight graduate students was assigned a different time series and each student used the three forecasting techniques on that assigned set of data. The usefulness of each technique was assessed by the root mean squared error (RMSE) of the students' one-step-ahead forecast. The lower the RMSE, the more accurate the forecast and hence the more useful the technique. Measure the agreement between RMSE values for the eight sets of time-series data in Table 6.7 and estimate their order of accuracy.

Table 6.7

Technique	RMSE values for series							
	1	2	3	4	5	6	7	8
Judgmental	7.8	8.6	6.8	1.0	1.5	1.3	3.0	3.6
Box–Jenkins	10.2	8.7	7.1	2.1	1.7	0.8	2.4	2.9
Exponential	8.5	14.0	10.5	1.8	1.3	1.3	2.7	4.7

The objects are the techniques and the series are the observers, so we have data to make $m = 8$ complete rankings of $n = 3$ objects. We rank from 1 to 3 for each series according to the RMSE values and sum the eight ranks, as shown in Table 6.8.

Table 6.8

Technique	Series								Sum
	1	2	3	4	5	6	7	8	
Judgmental	1	1	1	1	2	2.5	3	2	13.5
Box–Jenkins	3	2	2	3	3	1	1	1	16.0
Exponential	2	3	3	2	1	2.5	2	3	18.5
									48.0

The average rank is $8(4)/2 = 16.0$, the sum of squares of deviations is $S = 12.5$, and the coefficient of concordance from (6.4) is

$$W = \frac{12(12.5)}{64(24)} = 0.098. \tag{6.20}$$

If we make the correction for the two ties we have $U' = 0.5$ from (6.7) and from (6.8)

$$W = \frac{12.5}{[64(24)/12] - 8(0.5)} = 0.101.$$

The effect of the correction is small.

To test the significance of W, we go to Appendix Table 5 with $n = 3$, $m = 8$, $S = 12.5$. The exact probability is not found because of the ties, but we give the P-value as between 0.531 and 0.654 that correspond to $S = 14$ and $S = 8$ respectively. These data show no agreement between rankings and hence we should not estimate their order of accuracy.

Friedman test for randomised complete block designs

6.15 The previous test of significance for the null hypothesis of independence of m rankings of n objects can also be applied to the experimental design of observations on n treatments in m blocks. The null hypothesis is that the n treatment effects are all the same. The observations on the n treatments are ranked from 1 to n according to relative magnitude within each block separately to produce the m sets of rankings. The test based on S or z or χ_r^2 is then usually referred to as the Friedman test.

The true ranking of treatments in this design can be estimated after the manner of Section 6.12 according to the magnitude of the sums of ranks allotted to them. If r_s is calculated between this estimated ranking and each of the observed rankings, the average of these m calculated coefficients is a maximum. This estimated ranking is therefore the least squares estimate.

Example 6.7

Hartley (1973) examined the effects of noise on the performance of tasks by humans. Each subject faced five neon bulbs arranged in the form of a pentagon and five correspondingly arranged metal contacts. The subject was asked to touch the contact that corresponded to the lighted bulb. This action would light another bulb and the subject was then supposed to touch the contact corresponding to the newly lit bulb, and so on. In one part of the experiment the subjects were required to do the task for two 20-minute intervals. The design was a randomised complete block and the data given in the article were the mean numbers of errors made in each 5-minute block of time for the last 20-minute test. During each 20-minute interval the surroundings were either quiet or noisy in each possible combination. For example, *Q-N* indicates the first 20-minute test was under quiet conditions and the second test under noisy conditions. Use the data in Table 6.9 to test the null hypothesis that the four quiet–noise combinations of surroundings produce the same mean number of errors.

Table 6.9

Block	Quiet–noise combination			
	Q–Q	*Q–N*	*N–Q*	*N–N*
1	5.54	5.31	7.85	6.92
2	5.62	6.15	9.15	10.54
3	5.54	6.31	9.07	10.69
4	6.46	7.23	8.00	8.62

We have $n = 4$ treatments and $m = 4$ blocks. We rank the measures of error from 1 to 4 in each block, and sum the ranks for each treatment, as shown in Table 6.10.

Table 6.10

Q–Q	*Q–N*	*N–Q*	*N–N*
2	1	4	3
1	2	3	4
1	2	3	4
1	2	3	4
5	7	13	15

The sum of squares of deviations from (6.2) is $S = 68$ and $W = 0.018$ from (6.4). Appendix Table 5C shows that the *P*-value is 0.0027, so the different combinations of quiet and noise do tend to produce different error rates in task performance. The estimated ranking of treatments from smallest to largest number of errors is *Q-Q, Q-N, N-Q, N-N*.

Incomplete rankings

6.16 Cases sometimes occur in which the rankings are incomplete. One case of particular interest arises in experimental designs where we can deliberately arrange for them to be incomplete in a symmetrical way.

Consider, for example, a manufacturer of ice cream who wishes to test preferences for seven varieties by getting them ranked by a number of observers. He may not wish to present each of the varieties to each observer, either because of economy in time or because the human palate loses its discriminatory power after a few varieties are tasted. Calling the ice cream varieties A, B, C, D, E, F, G we then arrange them in sets of three as follows:

$$\begin{array}{ccccccc} A & B & C & D & E & F & G \\ B & C & D & E & F & G & A \\ D & E & F & G & A & B & C \end{array} \qquad (6.21)$$

One triplet of varieties will be presented to each of seven observers for ranking. The point of this arrangement is that every variety appears an equal number of times (three) in the entire experiment and each pair of objects is presented an equal number of times to each observer (once), so that all possible comparisons appear equally often. We can regard this as seven rankings of seven objects with three objects appearing at a time. This arrangement is called a Youden square; it combines the features of a balanced incomplete block design and a Latin square design.

6.17 Suppose the ranks given by the observers to the triplet of objects presented are as shown in Table 6.11.

Table 6.11

Observer						
1	2	3	4	5	6	7
A 2	B 2	C 1	D 1	E 2	F 2	G 3
B 1	C 3	D 2	E 3	F 3	G 3	A 1
D 3	E 1	F 3	G 2	A 1	B 1	C 2

We first rearrange the ranks according to object instead of observer and sum the ranks for each object as shown in Table 6.12.

Table 6.12

Object						
A	B	C	D	E	F	G
2	1	3	3	1	3	2
1	2	1	2	3	3	3
1	1	2	1	2	2	3
4	4	6	6	6	8	8

Notice that the total of all ranks is 42 or $(1 + 2 + 3)7$. If the ranks had been allotted at random by each observer, the sum of ranks for each object would tend to be equal to all other sums with a value of $42/7 = 6$. On the other hand, suppose a strict ordering of the objects exists and all observers agree upon that ordering, e.g. the best object would have only ranks of one. Then the sums of ranks would be some permutation of 3, 4, ..., 9, sums which are as different as possible.

As usual, we will let S denote the sum of squares of the deviations of the object sums in Table 6.12 from their mean of 6 to get

$$S = (4 - 6)^2 + (4 - 6)^2 + \cdots + (8 - 6)^2 = 16.$$

We construct a coefficient of agreement between rankings by dividing S by its maximum possible value, which here is 28, to get

$$W = 16/28 = 0.571.$$

6.18   In general, suppose we have n objects and the objects are presented k at a time so that each observer ranks only $k < n$ objects and each object is ranked a total of m times. Then the number of blocks of objects (observers) is mn/k, which must be an integer. Within each block each observer will make $k(k - 1)/2$ comparisons of pairs of objects for a total of

$$[k(k - 1)/2](mn/k) = mn(k - 1)/2$$

comparisons. Define λ as the number of blocks in which each specific comparison occurs. Since there are $n(n - 1)/2$ possible comparisons, the total number of comparisons is $\lambda n(n - 1)/2$. This must agree with the number of comparisons noted above. Therefore a restriction of this design is that

$$\lambda n(n - 1)/2 = mn(k - 1)/2$$

or

$$\lambda = \frac{m(k - 1)}{(n - 1)}. \tag{6.22}$$

Since λ must be an integer, (6.24) requires that $(n - 1)$ must be a factor of $m(k - 1)$.

In summary, the restrictions of this design are that $m(k - 1)/(n - 1)$ and mn/k must both be integers. Notice the change in notation from the description of the Friedman test in Section 6.15 where we had m blocks. The number of blocks here is mn/k.

Let R_i denote the sum of the ranks assigned to the ith object. The total of all ranks assigned to each object is $k(k + 1)/2$ and there are mn/k observers for a grand total $mn(k + 1)/2$. Hence the average rank sum per object is $m(k + 1)/2$ and the sum of squares of deviations is

$$S = \sum_{i=1}^{n} \left(R_i - \frac{m(k + 1)}{2} \right)^2 = \sum_{i=1}^{n} R_i^2 - \frac{nm^2(k + 1)^2}{4}. \tag{6.23}$$

The maximum value of S is attained when the object ranks sums are

$$m, m + \lambda, m + 2\lambda, \ldots, m + (n - 1)\lambda \tag{6.24}$$

and hence

$$\max S = \sum_{i=1}^{n} \left[m + (i - 1)\lambda - \frac{m(k + 1)}{2} \right]^2$$

$$= \sum_{i=1}^{n} \left[m + (i - 1)\lambda - \frac{m(k - 1)}{2} - m \right]^2$$

$$= \sum_{i=1}^{n} \left[(i - 1)\lambda - \frac{\lambda(n - 1)}{2} \right]^2$$

$$= \lambda^2 \sum_{i=1}^{n} \left(i - \frac{n + 1}{2} \right)^2 = \lambda^2 n(n^2 - 1)/12.$$

Thus in general the relative measure of agreement between rankings is

$$W = \frac{12S}{\lambda^2 n(n^2 - 1)}$$

$$= \frac{12 \sum_{i=1}^{n} R_i^2 - 3nm^2(k + 1)^2}{\lambda^2 n(n^2 - 1)}. \tag{6.25}$$

6.19 We may test the significance of W by taking (for moderate n and large m)

$$\chi^2 = \frac{\lambda(n^2 - 1)}{k + 1} W \tag{6.26}$$

to be distributed as χ^2 with $n - 1$ degrees of freedom. Appendix Table 8 is used to find P-values.

A better approximation is given by taking

$$z = \tfrac{1}{2} \log_e \frac{\left(\dfrac{\lambda(n + 1)}{k + 1} - 1 \right) W}{1 - W} \tag{6.27}$$

to be distributed as Fisher's z with

$$v_1 = \frac{mn \left(1 - \dfrac{k + 1}{\lambda(n + 1)} \right)}{\left(\dfrac{nm}{n - 1} - \dfrac{k}{k - 1} \right)} - \frac{2(k + 1)}{\lambda(n + 1)}, \tag{6.28}$$

$$v_2 = \left(\frac{\lambda(n + 1)}{k + 1} - 1 \right) v_1$$

degrees of freedom. Appendix Table 7A or 7B is used to obtain P-values.

If $k = n$ and hence $\lambda = m$, these reduce to equations (6.11) and (6.13) as they should.

   The advantages of this incomplete block design should be obvious. If every one of the n objects were ranked by each of the mn/k observers, a total of mn^2/k observations would be required. Here the total number of observations required is only mn. If the cost or time of the experiment is based on the total number of observations, as opposed to a cost per set of rankings, this design allows more objects to be compared for the same cost. Even more important, however, is the fact that this design generally produces more reliable results in the rankings, especially when the rankings are assigned by a subjective judgment. For example, in a tea-tasting experiment, it would be very difficult for even a professional tea taster to be able to rank say ten different varieties of tea with any feeling of confidence about the results.

Example 6.8

Durbin (1951) described an experiment where a design of type (6.21) was repeated three times to provide ranks by 21 observers of the $n = 7$ objects. The objects are still presented $k = 3$ at a time but now each object is ranked a total of $m = 9$ times so that $\lambda = 3$. The total of all ranks is $(1 + 2 + 3)21 = 126$ and the mean is $126/7 = 18$. The rank totals for the 7 objects are

A	B	C	D	E	F	G
20	13	18	25	22	12	16

The sum of squares of deviations is

$$S = (20 - 18)^2 + (13 - 18)^2 + \cdots + (16 - 18)^2 = 134$$

and the relative measure of agreement between rankings from (6.25) is

$$W = \frac{12(134)}{3^2(7)(48)} = 0.532.$$

   To test the significance of W with the chi-square approximation, we use (6.26) to find

$$\chi^2 = \frac{3(48)(0.532)}{4} = 19.15$$

with six degrees of freedom. Appendix Table 8 shows that the approximate P-value is less than 0.01 and we conclude that the preferences among varieties are not random.

   We use these same data to illustrate the Fisher z-approximation in (6.27) and (6.30). The test statistics is

$$z = \tfrac{1}{2} \log_e \frac{\left(\dfrac{3(8)}{4} - 1 \right) 0.532}{0.468} = 0.87$$

with

$$v_1 = \frac{9(7)\left(1 - \dfrac{4}{3(8)}\right)}{\dfrac{9(7)}{6} - \dfrac{3}{2}} - \frac{2(4)}{3(8)} = 5.50$$

$$v_2 = \left(\frac{3(8)}{4} - 1\right)(5.50) = 27.5.$$

Appendix Table 7B gives the following 1 per cent points of significance for z in the neighbourhoods of our degrees of freedom:

	$v_2 = 27$	$v_2 = 28$
$v_1 = 5$	0.6655	0.6614
$v_1 = 6$	0.6346	0.6303

Our $z = 0.87$ is clearly significant at the 1 per cent level so again we have P-value < 0.01 and the preferences are not random.

Notes and references

The primary references for the tests of $m > 2$ complete rankings given in this chapter are Friedman (1937, 1940), Kendall and Smith (1939) and Kendall (1942a).

Friedman (1940) published the significance points of S and this paper is the source of Appendix Table 6. More extensive tables have been developed by Owen (1962), Michaelis (1971), Quade (1972) and Odeh (1977) so that exact tables are available up to $n = 6$, $m = 6$. Sen and Krishnaiah (1984) also give tables. An approximation that is more accurate than that based on the chi-square distribution was developed by Jensen (1977) but it is computationally difficult. Iman and Davenport (1980) provided two new approximations which have good accuracy; one is very easily computed.

Noether (1987b) gave a method of determining the sample size required by the Friedman test to achieve a minimum specified power at a fixed level as a function of the probability of concordance.

Kraemer (1976) found that a function of the coefficient of concordance W is approximately distributed as F in the non-null case of a normal random effects model. She showed how to carry out hypothesis tests and find confidence interval estimates for the coefficient of reliability. Palachek and Schucany (1984) gave an approximate confidence interval for measures of concordance with a fixed set of objects.

Ehrenberg (1952) proposed the average of all possible values of Kendall tau as a measure of agreement that is preferable to the Kendall coefficient of concordance. Hays (1960) discussed this average. These authors found approximate asymptotic distributions for average tau. Later, Alvo, Cabillo and Feigin (1982) found its exact asymptotic distribution and showed that its Bahadur efficiency is larger than that of average rho. Alvo and Cabilio (1984) gave a further justification for using average tau, investigated the accuracy of approximations to its distribution, and gave tables of the exact distribution for small sample sizes. Alvo and Cabilio (1985) discussed average rank correlation statistics based on both tau and rho and in the presence of ties, and found the test based on average tau to be more efficient. Hubert (1979) developed a generalised concept of concordance that includes average rho and average tau but also allows extension to newer methods of data analysis in psychology.

The asymptotic relative efficiency of the Friedman test was investigated by Van Elteren and Noether (1959), Noether (1967), Sen (1967, 1972), Mehra and Sarangi (1967), Gilbert (1972), Silva and Quade (1983), Shirahata (1985), Agresti and Pendergast (1986), Ferretti and Yohai (1986), Yohai and Ferretti (1987), Hora and Iman (1988), Thompson and Ammann (1988), among others. Iman, Hora and Conover (1984) used a comprehensive Monte Carlo study to compare the small sample power and robustness of the Friedman test with some parametric and nonparametric competitors for five different distributions. The nonparametric competitors that performed well were the Quade (1979) test which is an extension of the Wilcoxon signed ranks test for paired samples and the Conover and Iman (1976) test which calculates the parametric *F*-test statistic from the ranks. Fawcett and Salter (1984) reported a similar study. Groggel (1987) compared the large sample approximations and power for a larger group of nonparametric tests that include the Friedman test.

Considerable research has been done on multiple comparisons procedures to determine which pairs of treatment differ in the randomised block design. We cite only Rosenthal and Ferguson (1965) and Wei (1982).

Buckley and Eagleson (1986) suggested using the largest of the Spearman rank correlation coefficients to determine which variables are associated and gave approximations to the distribution. Gordon (1979a, 1979b) proposed other measures of agreement between rankings that divide the objects into groups according to relative preferences and discussed the power of the corresponding tests.

As shown in Section 6.5, there is a linear relationship between the coefficient of concordance *W* and the average of all Spearman rho values. Taylor and Fong (1963) wrote about the computation and distribution of the average rho, and Taylor (1964) extended these methods to make corrections for both ties and continuity.

Jonckheere (1954a, 1954b) and Page (1963) developed tests based on average rank correlations for detecting ordered alternatives in randomised complete block designs. Hollander (1967), Skillings and Wolfe (1978), Salama and Quade (1981), Kepner and Robinson (1984) and Ferretti (1987) proposed related tests and made comparisons of efficiency.

The test described in this chapter for incomplete rankings is due to Durbin (1951). This test applies not only to Youden arrays but to incomplete block designs of all kinds (including paired-comparison designs). Designs of this type are given in Cochran and Cox (1957) and Kirk (1968). Paterson (1988) gives an essay about computer construction of these designs. Dykstra (1956) discussed applications. The efficiencies of the relevant tests were considered by Van Elteren and Noether (1959), Noether (1967), Salter and Fawcett (1985), among others. Van der Laan and Prakken (1972) and Skillings and Mack (1981) gave some tables of the null distribution. Fawcett and Salter (1987) investigated the adequacy of the chi-square approximation for the 58 designs given in Cochran and Cox (1957) and found it unsatisfactory in many cases. Van der Laan (1988) compared the accuracy of the chi-square and F-approximations for 11 designs.

A treatment of the general case when any number of ranks are missing from any ranking was given by Benard and Van Elteren (1953). This same paper extended the Durbin test to the case where replications are made on some experimental units. Brunden and Mohberg (1976) gave a simplified method for computing the test statistic.

The Friedman test has been extended to the case of replicated designs where each treatment is observed, say, k times in each block. The observations within each block are then ranked from 1 to kn.

Other nonparametric tests of the Friedman type have been proposed by Bhapkar (1961), Lemmer, Stoker and Reinach (1968), Downton (1976), Prentice (1979), Mack and Skillings (1980), Skillings and Mack (1981), De Kroon and Van der Laan (1983), Schemper (1984a), Groggel and Skillings (1986), Brits and Lemmer (1986), Burnett and Willan (1988), among others.

Quade (1984) gave a comprehensive overview with extensive references of nonparametric methods of the type covered in this chapter, with a good discussion of efficiency.

Problems

6.1   Juchau and Galvin (1984) reported a study in Australia to investigate whether the following three groups of accountants agree on the perceived importance of different kinds of communication skills needed by personnel:

    I.   corporate accountants;
  II.   chartered accountants;
 III.   academic accountants.

Questionnaires listing 20 representative communication skills were sent to staff partners of chartered accounting firms to obtain opinions for Group I, to chief accountants of listed public companies for Group II, and to heads of accounting departments in Australian colleges and universities for Group III. Respondents registered their opinions as to the extent to which each skill was needed using a six-point scale (1 = strongly disagree, 5 = strongly agree, 6 = can't evaluate). The average responses are shown in Table 6.13. Calculate the coefficient of concordance to measure the agreement between the three groups on the 20 communication skills.

Table 6.13

Communication skill	Corporate	Chartered	Academic
Use of visual aids	3.52	3.25	3.52
Reading speed	3.70	3.62	3.58
Paragraph development	3.95	4.16	4.00
Formal oral presentation	3.87	3.69	4.05
Deductive reasoning	4.31	4.19	4.26
Inductive reasoning	4.27	4.21	4.29
Correspondence writing	4.45	4.53	4.30
Correct punctuation	4.19	4.25	4.32
Correct grammar	4.34	4.48	4.32
Formal report writing	4.47	4.16	4.35
Conciseness	4.48	4.42	4.35
Listening responsiveness	4.39	4.51	4.36
Correct spelling	4.40	4.50	4.41
Memos and report writing	4.60	4.54	4.44
Listening attentiveness	4.59	4.55	4.44
Coherence	4.44	4.58	4.50
Outline development	4.27	4.21	4.54
Reading comprehension	4.66	4.63	4.58
Clarity	4.42	4.54	4.58
Informal oral presentation	4.23	4.47	4.58

6.2   Humphreys and Smith (1987) reported a study to investigate how well children are able to rank the physical strength of all the children in their class, including themselves. All children in a class were asked to rank the class members according to their perception of the children's physical strength. Consider the data in Table 6.14 for a sample of eight seven-year-old boys in a class and

(a)   use the Kendall coefficient of concordance to test the agreement between rankings and estimate the order of strength;
(b)   compute each student's average rank by all other students and find measures of association between those average ranks and each student's ranking of his own strength.

Table 6.14

		Student number							
		1	2	3	4	5	6	7	8
	1	5	1	8	3	7	4	2	6
	2	4	2	8	3	7	5	1	6
	3	4	2	5	3	8	6	1	7
Ranking	4	4	1	8	2	7	5	3	6
given by	5	4	1	8	3	5	6	2	7
student	6	5	1	8	3	7	4	2	6
	7	3	2	7	4	8	5	1	6
	8	4	1	8	2	6	5	3	7
	Sum	33	11	60	23	55	40	15	51

Answer: (a) $S = 2422$, $W = 0.901$, $\chi_7^2 = 50.46$, $P < 0.001$; $z = 2.077$, $v_1 = 6.75$, $v_2 = 47.25$, $P < 0.01$; Estimate order as 2, 7, 4, 1, 6, 8, 5, 3.

6.3 Goss and Karam (1987) described an experiment to evaluate the effects of three diets on the electrocardiographic responses during exercise. The diets used were mixed (M), high fat/high protein (HFHP) and high carbohydrate (HC). Each diet was especially formulated for each subject based upon carefully formulated guidelines and each subject's adherence to the assigned diet was carefully monitored. A randomised complete block design was used for the three diets and six male subjects. At the completion of each diet, each subject ran on a level treadmill until exhausted and the time to exhaustion was measured in minutes to produce the data in Table 6.15. Use the Friedman test to determine whether the diets have the same effect on endurance as measured by time to exhaustion.

Table 6.15

	Diet		
Subject	M	HFHP	HC
1	84	91	122
2	35	48	53
3	91	71	110
4	57	45	71
5	56	61	91
6	45	61	122

Answer: $S = 56$, P-value $= 0.006$.

6.4 Russell and Taylor (1985) used a simulation analysis of a hypothetical assembly to evaluate 11 sequencing rules, some common to simple job shops and some designed specifically for assembly shops. The performance of the sequencing rules was evaluated with respect to five characteristics relating to efficiency for ten simulation runs of 500 completed jobs for each sequencing rule. The summary statistics are shown in Table 6.16, where smaller numbers indicate better performance. Calculate the coefficient of concordance to measure the agreement between rankings of the 11 sequencing rules for the five characteristics. Which sequencing rule is best?

6.5 Howell and Johnson (1982) reported an experiment to compare the responses of students to the format of introductory and intermediate level accounting courses at Rice University in Texas. Several sections of each course were offered in the regular semester format and under a

Table 6.16

Sequencing rule	Characteristic				
	Mean flow time	Mean tardiness	Percent tardy	RMS tardiness	Mean assembly delay
FISFS	3.49	2.58	0.25	37.31	0.82
SPT	6.05	7.62	0.21	132.70	4.37
DDATE	2.50	0.44	0.009	1.71	0.86
RWK	2.51	3.26	0.017	14.34	0.95
(ROPT)²	2.50	4.58	0.01	11.41	0.87
SP	5.46	8.55	0.13	102.19	4.15
BS	6.30	4.29	0.32	67.68	1.05
RWK + SC	3.46	4.67	0.06	42.77	2.05
SC + (ROPT)²	2.53	8.54	0.01	28.28	0.97
BS + (ROPT)²	2.81	1.21	0.02	5.93	0.76
LP + (ROPT)²	2.37	2.58	0.01	10.79	0.64

Answer: $W = 0.747$, $\chi^2_{10} = 37.04$, DDATE is best.

compressed format (three weeks with three hours of class time daily) over a period of two years. All extraneous variables were controlled as rigidly as possible for each level of course. One instructor taught all sections of each level of course using the same text and syllabus; problems and tests were as similar as possible without being identical. At the end of each course the students used a standard evaluation instrument that covered 14 aspects of the course. Two aspects dealt with grades. The remaining 12 aspects were ratings of characteristics of the course on a ten-point scale. The average ratings on these 12 aspects are shown in Table 6.17 for the four combinations of course level and format:

CC1 = compressed course introductory;
CC2 = compressed course intermediate;
R1 = regular semester introductory;
R2 = regular semester intermediate.

(a) Compare the effectiveness in all four types of classes.
(b) Since CC1 and R1 were taught by one instructor and CC2 and R2 were taught by a different instructor, measure separately for each instructor the agreement between rankings between the compressed and regular formats.

6.6 Lusch and Ross (1985) investigated whether the power relationships of pairs of food brokers (*A*) and wholesalers (*B*) were limited in scope. The analysis centred on (1) *A*'s self-perception of power over *B*, and (2) *B*'s self-perception of power over *A*. Their research hypotheses were that both would be limited in scope. A sample of 54 matched pairs of *A* and *B* was obtained and each person was questioned regarding the following

Table 6.17

	CC1	CC2	R1	R2
Amount of material covered	6.9	7.1	7.4	7.7
Effort required	7.7	8.7	8.1	8.3
Personal attention received	6.5	7.4	6.3	6.8
Thoroughness	7.1	6.9	8.6	8.2
Student motivation	8.1	7.5	7.0	8.2
Own capability	6.5	6.5	7.4	7.1
Effectiveness of format	7.7	6.0	7.6	6.3
Amount of time required	7.5	8.6	7.5	8.1
Own attendance	8.4	9.0	9.2	9.0
Class performance	7.1	7.1	7.1	7.1
Course importance	7.3	7.9	7.5	7.2
Own motivation	8.1	8.7	8.0	8.8

nine marketing policy areas:

- I.   depth of inventory
- II.   size of inventory
- III.   price level
- IV.   amount of sales promotion
- V.   product mix
- VI.   product addition
- VII.   order size
- VIII.   decisions to delete product
- IX.   choice of transportation made.

Each A member of the pair was asked to use a rating scale of 1 to 5 for each of these nine areas where $1 = $ I have considerably less influence than B and $5 = $ I have considerably more influence than B. Each B member had the same rating scale for his influence relative to A. Each sample member's ratings were then ranked from 1 to 9 for the nine policy areas to obtain results like the following for A:

	I	II	III	IV	V	VI	VII	VIII	IX
1	1	2	5	3.5	7.5	7.5	7.5	7.5	3.5
2	1.5	3	1.5	6.5	4.5	8.5	6.5	8.5	4.5
⋮									
54	3.5	1	2	5	3.5	7.5	7.5	7.5	7.5
$\sum Rj$	314	305	323	213	272	238	224	273	268

and similarly for B:

$\sum Rj$	287	285	346	225	285	228	270	315	189

Calculate the coefficient of concordance to measure the agreement between the 54 ratings given by sample members

(a) A

(b) B

without using the correction for ties and test the null hypothesis of no agreement.

Answers: (a) $W = 0.071$, $\chi_8^2 = 30.6$, *P*-value < 0.001
(b) $W = 0.108$, $\chi_8^2 = 46.6$, *P*-value < 0.001.

6.7 Sheehan, Rowland and Burke (1987) reported a study to compare the effectiveness of four different treadmill protocols in eliciting maximal oxygen consumption (VO_2 max) when testing boys aged 10 to 12 years. Many physiologists use VO_2 max as an indication of cardiovascular fitness. The four treadmill protocols were (1) the continuous walking machine, (2) the continuous running machine, (3) the intermittent running machine and (4) the continuous running machine while holding the handrails. Sixteen boys aged 10–12 were tested for VO_2 max with each protocol during each of four different weeks. The data below represent the total VO_2 max scores for the 16 boys in each of the protocol–week combinations. Test the null hypothesis that the median VO_2 scores are the same for the four protocols.

Table 6.18

	Protocol			
Week	1	2	3	4
1	22	26	24	21
2	25	27	22	20
3	24	20	21	23
4	21	22	23	24

Answer: $S = 2$, $P = 0.992$.

6.8 Sands, Newby and Greenberg (1981) tried to determine whether there is a relationship between level of identified health risk in the work setting and illness behaviour as measured by visits to the company dispensary. After a physician at the B. F. Goodrich chemical plant discovered that vinyl chloride was associated with the development of a rare liver disease, all employees were offered regular medical screening at the dispensary. All industrial workers were classified into three risk classes: (1) low risk (negative test results), (2) moderate risk (some abnormal test results), (3) high risk (positive test results that required a change in job environment). Members of the three risk classes were then matched by sex, age and type of exposure, and samples of three matched risk groups of 24 workers each were chosen for observation for two time periods. Determine from Table 6.19 whether the median number of visits to the dispensary is the same for the three risk groups.

Table 6.19

Worker code	Frequency of visits		
	High risk	Low risk	Moderate risk
1	1	4	3
2	0	11	10
3	20	5	14
4	0	4	14
5	12	0	4
6	5	7	10
7	9	1	24
8	28	1	14
9	19	1	7
10	27	9	11
11	17	2	2
12	3	4	2
13	2	1	5
14	8	3	1
15	8	5	9
16	3	11	7
17	1	1	2
18	6	0	0
19	8	1	6
20	6	5	23
21	3	5	3
22	7	22	12
23	3	15	5
24	3	33	1

Answer: $\chi_2^2 = 0.8125$, $0.50 < P$-value < 0.70.

6.9 Rickey (1954) gave data on five variables for the 'greatest pitchers' in baseball since 1920. The five variables were (1) hits divided by at bats, (2) bases on balls plus hit batsmen divided by the sum of at bats, bases on balls, and hit batsmen, (3) earned runs divided by the sum of hits, bases on balls, and hit batsmen, (4) strike-outs divided by the sum of at bats, bases on balls, and hit batsmen, (5) earned runs per nine innings. The figures below are percentages for these five variables. Determine whether there is a significant agreement among rankings of players for the variables.

Table 6.20

Pitcher	(1)	(2)	(3)	(4)	(5)
Carl Hubbell	25.1	5.3	28.0	11.2	2.98
Dizzy Dean	25.3	6.0	27.6	14.4	3.03
Lefty Grove	25.4	7.5	26.7	13.6	3.09
Grover Alexander	27.3	3.8	28.5	6.4	3.09
Dazzy Vance	25.4	7.6	28.6	16.8	3.22
Dutch Leonard	26.5	6.0	28.4	8.8	3.25
Bucky Walters	25.4	9.0	27.4	8.8	3.30
Walter Johnson	25.6	8.0	28.6	12.0	3.33
Lefty Gomez	24.3	10.5	27.3	13.6	3.34
Paul Derringer	27.2	5.2	29.8	9.6	3.46
Fred Fitzsimmons	27.2	6.7	29.8	6.4	3.54
Ted Lyons	27.5	6.6	30.1	6.4	3.67

6.10   Elbert and Anderson (1984) studied student reaction to three different kinds of cases used in teaching marketing management in colleges of business administration. The cases used represented those randomly selected directly from the textbook, local cases drawn from area businesses and circulated to students as text supplements, and non-local cases circulated as text supplements. Each student evaluated the cases on a five-point Likert scale (5 = strongly agree) with respect to nine criteria statements. The group means are shown in Table 6.21. Determine whether the mean scores are the same for the three types of cases.

Table 6.21

Statement	Text	Local	Non-local
1. Case helped me to understand concepts in textbook	3.78	3.84	3.48
2. Case helped me to analyse problems	3.92	4.01	3.73
3. Case helped me to learn to make decisions	3.75	3.93	3.56
4. Case helped me know about real world	4.15	4.28	3.79
5. Case helped me develop skill in expressing ideas	3.70	3.81	3.65
6. Case was interesting	3.90	4.05	3.45
7. Topic of case was relevant to my curriculum	3.83	3.93	3.85
8. Case dealt with a timely business issue	4.18	4.32	3.79
9. Case was worthwhile overall	3.96	4.12	3.73

6.11   Muczyk and Gable (1981) reported a study where five top managers of a large greeting card store chain ranked the importance (1 = most important) of six performance dimensions to successful management. It is important that the rankings show a significant agreement because the authors went on to use the average ranks as weights for evaluations of 27 store managers. Test the agreement using the data in Table 6.22. The performance dimensions are AD-ability to deal with customer, AW-ability to work with subordinates, CC-carrying out company policy, MS-merchandising skill, PK-product knowledge, and RA-routine administrative functions.

Table 6.22

Top manager	Dimension					
	AD	AW	CC	MS	PK	RA
AA	5	4	2	1	3	6
BB	5	4	3	2	1	6
CC	2	3	1	4	5	6
DD	5	6	4	1	2	3
EE	5	4	1	6	2	3

6.12   Robichaud and Wilson (1976) reported a study of college undergraduates divided into three groups, prejudiced whites (PW), unprejudiced whites (UW) and blacks (B). Each student completed a questionnaire to rank 28 possible goals of social action on a five-point Likert scale (1 = top priority). The 28 goals were then grouped into five categories labelled Rights, Integration, Self-help, Hand-out and Right to Integrate, and assigned rankings from 1 to 5 according to the student rankings of importance of the individual goals within each group to obtain the data in Table 6.23. Determine whether there is significant agreement between rankings.

Table 6.23

	Goal type				
	Rights	Integration	Self-help	Hand-out	Right to integrate
PW	1	5	2	3	4
UW	1	4	2	5	3
B	1	5	2	4	3

Answer: $S = 80$, $m = 3$, $n = 5$, P-value 0.004.

6.13 Lawrence and Steed (1984) gathered information about types of student disruptive behaviour causing concern to teachers and administrators in five different European countries. The paper gave the overall summary rankings below for 15 types of behaviour where 1 = least concern and 15 = concern. Determine whether there is a significant association among rankings by country.

Table 6.24

Behaviour	France	W. Germany	Switzerland	Denmark	England
1. Bullying or physical violence to other children	2	1	10	2	1
2. Vandalism	1	2	4	6	4
3. Rowdy behaviour	8	3	10	6	13
4. Truancy	10.5	4.5	3	1	6
5. Refusal to obey	3.5	4.5	7	3	3
6. Difficult classes	3.5	6.5	1	6	5
7. Boredom	10.5	6.5	5.5	11	7
8. Temper	10.5	8.5	12	9	12
9. Verbal abuse or bad language	6	8.5	2	4	2
10. Extreme tardiness	13	10.5	10	13	11
11. Alcoholism	14	10.5	13	13	15
12. Talking/chatting	15	12	5.5	9	8
13. Difficult schools	10.5	13	14	13	10
14. Stealing	6	14	8	9	9
15. Physical violence to teachers	6	15	15	15	14

Answer: $S = 4593.5$, $\chi_{14}^2 = 45.9$, P-value < 0.01.

Chapter 7

Proof of the results of Chapter 6

7.1 We shall first establish the validity of using Fisher's z-distribution to provide an approximate test of the concordance coefficient W in the population of $(n!)^m$ possible rankings. As we shall require general results for tied rankings, we give a general investigation, due to Pitman (1938).

7.2 Suppose we have m sets of numbers

$$
\begin{array}{cccc}
a_1 & a_2 & \cdots & a_n \\
b_1 & b_2 & \cdots & b_n \\
\vdots & \vdots & & \vdots \\
k_1 & k_2 & \cdots & k_n.
\end{array}
\tag{7.1}
$$

We will suppose that each is measured about the mean of the row in which it occurs, so that the means of rows and the mean of the whole are zero. We then have

$$
W = \frac{(1/m) \sum_1^n (a_j + b_j + \cdots + k_j)^2}{\sum_1^n a_j^2 + \sum_1^n b_j^2 + \cdots + \sum_1^n k_j^2}.
\tag{7.2}
$$

The denominator in the expression is a constant and the variability arises solely from the numerator. If α_2 represents the second moment of the a-row and so on for b, c, etc., the denominator is $n \sum \alpha_2$. Writing

$$
R_{ab} = \sum_1^n a_i b_i
\tag{7.3}
$$

and

$$
U = \sum_1^{\binom{m}{2}} R_i,
\tag{7.4}
$$

where R_i stands generally for any R_{ab}, we have

$$
W = \frac{1}{m} + \frac{2U}{mn \sum \alpha_2}.
\tag{7.5}
$$

We find first the moments of R_{ab}, then those of U and finally those of W.

7.3 The expected values are

$$
E(R_{ab}) = 0
$$

$$E(R_{ab}^2) = E(\sum a_i b_i)^2$$

$$= E(\sum a_i^2 b_i^2 + \sum{}' a_i b_i a_j b_j)$$

$$= nE(a_i^2 b_i^2) + n(n-1)E(a_i a_j b_i b_j)$$

$$= nE(a_i^2)E(b_i^2) + \frac{n(n-1)}{n^2(n-1)^2}[E(\sum a_i)^2 - E\sum a_i^2]$$

$$\times [E(\sum b_i)^2 - E\sum b_i^2]$$

$$= n\alpha_2\beta_2 + \frac{n}{n-1}\alpha_2\beta_2 = \frac{n^2\alpha_2\beta_2}{n-1}. \qquad (7.6)$$

In a similar way—we omit the algebraical details—we find

$$E(R_{ab}^3) = \frac{n^3\alpha_3\beta_3}{(n-1)(n-2)} = \frac{(n-1)(n-2)}{n}\alpha_3'\beta_3', \qquad (7.7)$$

$$E(R_{ab}^4) = \frac{3n^4\alpha_2^2\beta_2^2}{(n-1)(n+1)} + \frac{(n-1)(n-2)(n-3)}{n(n+1)}\alpha_4'\beta_4', \qquad (7.8)$$

where we write α_3 for the third moment of the a-array and α_3', α_4' for its third and fourth k-statistics, the unbiased estimators of the cumulants of the distribution, which are defined in terms of the moments by

$$\alpha_3' = \frac{n^2}{(n-1)(n-2)}\alpha_3$$

$$\alpha_4' = \frac{n}{(n-1)(n-2)(n-3)}[(n+1)\alpha_4 - 3(n-1)\alpha_2^2].$$

The point of using k-statistics is that for normal populations they vanish for degree higher than 2 and may therefore be presumed small for populations reasonably close to normality.

7.4   To find the moments of U we note that

$$E(R_{ij}R_{kl}) = 0 \quad \text{for all suffixes except } i = k, j = l$$

$$E(R_{ij}R_{kl}R_{mn}) = 0 \quad \text{unless the suffixes form a 'circular' set such as } ij, jk, ki.$$

Similarly with four terms the only type not vanishing is one in which the suffixes are ij, jk, kl, li. Hence

$$E(U) = E(\sum R) = 0 \qquad (7.9)$$

$$E(U^2) = E(\sum R_i^2 + \sum{}' R_i R_j)$$

$$= \frac{n^2}{n-1}\sum \alpha_2\beta_2. \qquad (7.10)$$

Further

$$E(R_{ij}R_{jk}R_{ki}) = E(\sum a_i b_i \sum b_i c_i \sum c_i a_i)$$

$$= E\{[\sum b_i^2 a_i c_i + \sum' b_i b_j(a_i c_j + a_j c_i)] \sum c_i a_i\}$$

$$= E[Eb_i^2 \sum a_i c_i + Eb_i b_j \sum (a_i c_j + a_j c_i)] \sum c_i a_i$$

$$= (Eb_i^2 - Eb_i b_j)E(\sum a_i c_i)^2$$

$$= \left(\beta_2 + \frac{\beta_2}{n-1}\right) \frac{n^2 \alpha_2 \gamma_2}{n-1}$$

$$= \frac{n^3 \alpha_2 \beta_2 \gamma_2}{(n-1)^2}$$

$$\sum E(R_i^3) = \frac{(n-1)(n-2)}{n} \sum \alpha_3' \beta_3'.$$

Hence

$$E(U^3) = \frac{6n^3}{(n-1)^2} \sum \alpha_2 \beta_2 \gamma_2 + \frac{(n-1)(n-2)}{n} \sum \alpha_3' \beta_3'. \tag{7.11}$$

Finally—again we omit the algebra:

$$E(U^4) = \frac{3n^4}{(n-1)(n+1)} \sum \alpha_2^2 \beta_2^2 + \frac{(n-1)(n-2)(n-3)}{n(n+1)} \sum \alpha_4' \beta_4'$$

$$+ \frac{3n^4}{(n-1)^2} \{(\sum \alpha_2 \beta_2)^2 - \sum \alpha_2^2 \beta_2^2\}$$

$$+ 12(n-2) \sum \alpha_3' \beta_3' \gamma_2 + \frac{72n^4}{(n-1)^3} \sum \alpha_2 \beta_2 \gamma_2 \delta_2. \tag{7.12}$$

Finally, for the moments of W,

$$E(W) = \frac{1}{m} = \overline{W}, \quad \text{say} \tag{7.13}$$

$$E(W - \overline{W})^2 = \frac{4}{m^2(n-1)} \frac{\sum \alpha_2 \beta_2}{(\sum \alpha_2)^2}, \tag{7.14}$$

and, neglecting terms involving α_3' and α_4' which are of lower order in n,

$$E(W - \overline{W})^3 = \frac{48}{m^3(n-1)^2} \frac{\sum \alpha_2 \beta_2 \gamma_2}{(\sum \alpha_2)^3} \tag{7.15}$$

$$E(W - \overline{W})^4 = \frac{48}{m^4(n-1)^2} \frac{(\sum \alpha_2 \beta_2)^2}{(\sum \alpha_2)^4} - \frac{96}{m^4(n-1)^2(n+1)} \frac{\sum \alpha_2^2 \beta_2^2}{(\sum \alpha_2)^4}$$

$$+ \frac{1152}{m^4(n-1)^3} \frac{\sum \alpha_2 \beta_2 \gamma_2 \delta_2}{(\sum \alpha_2)^4}. \tag{7.16}$$

7.5   Consider now the case when the ranks are untied. Then all the variances of rankings are equal and we find

$$E(W) = \frac{1}{m} \tag{7.17}$$

$$\mu_2(W) = E(W - \overline{W})^2 = \frac{2(m - 1)}{m^3(n - 1)} \tag{7.18}$$

$$\mu_3(W) = E(W - \overline{W})^3 = \frac{8(m - 1)(m - 2)}{m^5(n - 1)^2} \tag{7.19}$$

$$\mu_4(W) = E(W - \overline{W})^4 = \frac{12(m - 1)^2}{m^6(n - 1)^2} + \frac{48(m - 1)(m - 2)(m - 3)}{m^7(n - 1)^3}$$

$$- \frac{48(m - 1)}{m^7(n - 1)^2(n + 1)}. \tag{7.20}$$

Now consider the Beta or Type I distribution

$$dF = \frac{1}{B(p, q)} W^{p-1}(1 - W)^{q-1} dW. \tag{7.21}$$

The first two moments are

$$E(W) = \frac{p}{p + q} \tag{7.22}$$

$$\mu_2(W) = \frac{pq}{(p + q)^2(p + q + 1)}. \tag{7.23}$$

If we equate (7.17) to (7.22) and (7.18) to (7.23) we find

$$p = \tfrac{1}{2}(n - 1) - \frac{1}{m}$$

$$q = (m - 1)p. \tag{7.24}$$

The distribution of W thus can approximate (7.21) with p and q given by (7.24). The third moment of (7.21).

$$\frac{8(m - 1)(m - 2)}{m^4(n - 1)(mn - m + 2)} = \frac{8(m - 1)(m - 2)}{m^5(n - 1)^2}\left[1 - \frac{2}{m(n - 1) + 2}\right].$$

Comparing this with (7.19) we see that the third moment of (7.21) is approximately equal to the third moment of W unless $m(n - 1)$ is small. Again the fourth moment of (7.21) is

$$\frac{12(m - 1)}{m^6(n - 1)^2}\left[\frac{(m - 1)y^2 + 4m^2y - 14(m - 1)y}{(y + 2)(y + 4)}\right],$$

where $y = m(n - 1)$, and this is approximately equal to the fourth moment of W in (7.20) if $m(n - 1)$ is not too small.

7.6   The first two moments of the distribution in (7.21) are exactly equal to the first two moments of W, and the third and fourth moments of (7.21) are approximately equal to the corresponding moments of W; and thus we expect (7.21) to provide an adequate approximation. The accuracy of the approximation is, in fact, greater than perhaps our rather long proof of the result might foreshadow.

A simple transformation reduces the Beta distribution to the form of Fisher's z-distribution. If we substitute

$$z = \tfrac{1}{2} \log_e \frac{(m-1)W}{1-W}$$

in (7.21), the density becomes

$$dF \propto \frac{e^{2pz}\, dz}{\{(m-1) + e^{2z}\}^{p+q}}$$

which is Fisher's form with

$$v_1 = 2p$$

$$v_2 = 2q$$

so that, from (7.24),

$$v_1 = (n-1) - \frac{2}{m}$$

$$v_2 = (m-1)v_1$$

as given in (6.12).

7.7   If ties are present the test needs further consideration.

(a)   In the above derivation we have only used the absence of ties to evaluate the terms α_2, and if all rankings are equally tied these variances are still equal and the results hold.

(b)   If the U'-numbers appropriate to ties are small compared with $\frac{1}{12}(n^3 - n) = N$, say, again, the test requires no modification. For then, to the first order in $\sum U'/mN$,

$$\frac{\sum \alpha_2 \beta_2}{(\sum \alpha_2)^2} = \frac{\sum (N - U'_\alpha)(N - U'_\beta)}{\sum (N - U'_\alpha)^2}$$

$$= \frac{(m-1)m}{m^2}\left(1 - \frac{2}{mN}\sum U'\right)\left(1 - \frac{2}{mN}\sum U'\right)^{-1}$$

$$= \frac{m-1}{m}$$

so that, to this order, the second moment of W remains unchanged. The effect on the third and fourth moments is also negligible to this order. Hence our result.

(c)   If the U'-numbers are large then we must calculate $\mu_2(W)$. We shall then find

$$\tfrac{1}{2}v_1 = p = \frac{m-1}{m^3\mu_2(W)} - \frac{1}{m} \tag{7.25}$$

$$\tfrac{1}{2}v_2 = q = (m-1)p$$

and the test may be applied with these values of v_1 and v_2.

7.8   We now prove that the statistic

$$\chi_r^2 = m(n-1)W$$

$$= \frac{S}{\tfrac{1}{12}mn(n+1)} \tag{7.26}$$

tends as m increases to that of the χ^2-distribution

$$\mathrm{d}F \propto e^{-(1/2)\chi^2}\chi^{v-1}\,\mathrm{d}\chi \tag{7.27}$$

with $v = n - 1$.

In the array (7.1) consider the sum of any column, say the first, which we will call p. We have

$$E(p) = E(a_1 + b_1 + \cdots + k_1) = 0$$

$$E(p^2) = E(\textstyle\sum a)^2 \qquad\qquad = \sum \alpha_2$$

$E(p^{2r+1})$ is of order $\sum \alpha_{2r+1} + \sum \alpha_{2r-1}\beta_2 + \cdots$. $E(p^{2r})$ is of order $\sum \alpha_{2r} + \cdots + \sum \alpha_2\beta_2 \cdots \kappa_2$. The same argument as we employed in Section 5.21 leads to the conclusion that $E(p^{2r+1})$ is of lower order in m and that the dominant term in $E(p^{2r})$ is $\sum \alpha_2\beta_2 \cdots \kappa_2$. Thus in the limit odd moments vanish and

$$\frac{\mu_{2r}}{(\mu_2)^r} \sim \frac{(2r)!}{2^r} \frac{\sum \alpha_2\beta_2 \cdots \kappa_2}{(\sum \alpha_2)^r}. \tag{7.28}$$

When all αs are equal, or nearly so,

$$\frac{\mu_{2r}}{(\mu_2)^r} \sim \frac{(2r)!}{r!\,2^r}$$

and hence the distribution of p tends to normality with zero mean.

Now S is the sum of squares of n such variates, subject only to the constraint that $\sum p = 0$. Thus kS is distributed as χ^2 with $v = n - 1$, where k is a factor to be determined such that the mean of kS is $n - 1$, the mean of χ^2. But

$$E(kS) = kn \sum \alpha_2$$

and thus

$$k = \frac{n-1}{n \sum \alpha_2}$$

and hence

$$\chi_r^2 = \frac{(n-1)S}{n \sum \alpha_2}$$

is distributed as χ^2.

When the rankings contain no ties each has the variance $\frac{1}{12}(n^2 - 1)$ and thus

$$\chi_r^2 = \frac{S}{\frac{1}{12}nm(n+1)}$$

as given in (7.26).

If ties are present, represented by U'-numbers, the appropriate value is

$$\chi_r^2 = \frac{(n-1)S}{n[\frac{1}{12}m(n^2 - 1) - (1/n) \sum U']}$$

$$= \frac{S}{\frac{1}{12}mn(n+1) - [1/(n-1)] \sum U'}. \qquad (7.29)$$

Unless the ties are substantial in extent the effect of the second term in the denominator is small.

7.9   We now indicate the basis of calculating the actual distribution of W (or equivalently of S) for smaller values of m and n.

For $m = 2$ the values are derivable from the distribution of Spearman's rho. We proceed from the case for given m, n to that for $m + 1, n$. For example, with $m = 2, n = 3$, we have the values shown in Table 7.1 for the sums of ranks measured about their mean.

Table 7.1

Type			Frequency
−2	0	2	1
−2	1	1	2
−1	0	1	2
0	0	0	1

Here $-2, 1, 1$ and $2, -1, -1$ are taken to be identical types, for they give the same value of S and will also give similar types when we proceed to the case $m = 3$ as follows.

For $m = 3$ each of the above types will appear added to six permutations of $-1, 0, 1$; e.g. the type $-2, 0, 2$ will give one each of $-3, 0, 3$; $-3, 1, 2$; $-2, -1, 3$; $-2, 1, 1$; $-1, -1, 2$ and $-1, 0, 1$. These types are counted for each of the basic types of $m = 2$, and we get Table 7.2.

For $n = 5$ and greater, the labour becomes very considerable owing to the large number of different types to be taken into account at each stage. It seems, however, that for all ordinary purposes in testing significance the z-distribution provides an adequate approximation for larger values of n.

Table 7.2

Type			Frequency
−3	0	3	1
−3	1	2	6
−2	0	2	6
−2	1	1	6
−1	0	1	15
0	0	0	2
			36

7.10   We have now to show (as indicated in Section 6.12) that the method of estimation there proposed is such as to maximise the average Spearman's rho between the estimated and the observed rankings.

Suppose the estimated ranking is $X_1, \ldots, X_n$ and let the sums of ranks be $S_1, \ldots, S_n$. Then the average rho is given by

$$\frac{1}{m} \sum_{k=1}^{m} r_{sk} = \frac{12}{m(n^3 - n)} \sum_{k=1}^{m} \sum_{j=1}^{n} \{X_j - \tfrac{1}{2}(n+1)\}\{x_{jk} - \tfrac{1}{2}(n+1)\},$$

where x_{jk} is the rank of the kth object in the jth ranking. This is equal to

$$\frac{12}{m(n^3 - n)} \sum_{j=1}^{n} \{X_j - \tfrac{1}{2}(n+1)\}\{S_j - \tfrac{1}{2}m(n+1)\}$$

$$= \frac{12}{m(n^3 - n)} \left[\sum_{j=1}^{n} (X_j S_j) - \tfrac{1}{4}mn(n+1)^2 \right]. \tag{7.30}$$

This is clearly a maximum when $\Sigma(XS)$ is a maximum, i.e. when the greatest S is multiplied by the greatest X and so on, the least S being multiplied by the least X. Our suggested rule of estimation does in fact ensure that the multiplications take place in this way, and hence the result follows.

If we consider the sum

$$U = \sum_{j=1}^{n} (S_j - mX_j)^2$$

$$= \Sigma S_j^2 + m^2 \Sigma X^2 - 2m \Sigma(XS),$$

we see that, since the first two terms on the right are constants, U is minimised if $\Sigma(XS)$ is maximised. Our method of estimation therefore minimises U, that is to say, minimises the sum of squares of differences between the actual sums S and what they would be, mX, if all rankings were identical.

7.11   We now establish the test for the case of incomplete rankings considered in Section 6.16. With the notation there used let us put

$$U = \sum_{j} \sum_{i<p} x_{ij} x_{pj}. \tag{7.31}$$

Then, since

$$\sum x_{ij}^2 = \frac{mn(k^2 - 1)}{12},$$

we find

$$W = \frac{12}{\lambda^2 n(n^2 - 1)} \left(\sum x_{ij}^2 + 2 \sum_j \sum_{i<p} x_{ij} x_{pj} \right)$$

$$= \frac{k + 1}{\lambda(n + 1)} + \frac{24U}{\lambda^2 n(n^2 - 1)}. \tag{7.32}$$

Now the jth object occurs in different blocks in the ith and pth replication, and an x in any one block is independent of an x in any other block. Hence

$$E(U) = 0 \tag{7.33}$$

and

$$U^2 = \sum x_{ij}^2 x_{pj}^2 + 2 \sum x_{ij} x_{pj} x_{ql} x_{rl}. \tag{7.34}$$

There are $\frac{1}{2} nm(m - 1)$ terms in the first of these sums and the expectation of each is $\{E(x_{ij}^2)\}^2 = \frac{1}{144}(k^2 - 1)^2$. The only terms contributing to the second sum are those for which the jth and lth objects occur together in different blocks. Now

$$E(x_{ij} x_{ql}) = E(x_{pj} x_{rl})$$

$$= -\tfrac{1}{12}(k + 1),$$

when the jth and lth object occur in the same block, and is zero in the contrary case. The number of such cases of occurrence together is $\frac{1}{4} n(n - 1)\lambda(\lambda - 1)$. Hence, substituting in (7.34) we get

$$E(U^2) = \frac{mn(k + 1)(k^2 - 1)}{288} [(m - 1)(k - 1) + (\lambda - 1)]. \tag{7.35}$$

Thus

$$E(W) = \frac{k + 1}{\lambda(n + 1)}, \tag{7.36}$$

$$\text{var } W = \frac{2(k + 1)^2}{mn\lambda^2(n + 1)^2} \left(m - 1 + \frac{\lambda - 1}{k - 1} \right). \tag{7.37}$$

7.12   It is to be noted that we cannot proceed to calculate third and fourth moments in the way employed for the ordinary coefficient because the symmetries holding by virtue of the equal occurrence of objects and pairs of objects do not hold for triads or more complex sets. Arguing (a little heuristically) by analogy from W in the ordinary case we may, however, identify our distribution with that of (7.21) by equating its first and second moments to

(7.36) and (7.37) respectively. This gives us

$$p = \frac{mn\left(1 - \dfrac{k + 1}{\lambda(n - 1)}\right)}{2\left(\dfrac{mn}{n - 1} - \dfrac{k}{k - 1}\right)} - \frac{k + 1}{\lambda(n + 1)} \tag{7.38}$$

$$q = \left(\frac{\lambda(n - 1)}{k + 1} - 1\right)p$$

and the test of Section 6.19 follows.

7.13 To fit a χ^2 distribution, which has $n - 1$ degrees of freedom, we put $W = a\chi^2$ and find

$$E(W) = \frac{k + 1}{\lambda(n + 1)} = a(n - 1)$$

and hence, evaluating a, we see that

$$\chi^2 = \frac{\lambda(n^2 - 1)}{k + 1} W \tag{7.39}$$

tends to be distributed as χ^2 with $n - 1$ degrees of freedom. The variance of this χ^2 is

$$2(n - 1)\left\{1 - \frac{k(n - 1)}{mn(k - 1)}\right\} = 2(n - 1)\left(1 - \frac{1}{m}\right) \tag{7.40}$$

approximately, and this agrees with (7.37) for large m.

Notes and references

See the notes and references to Chapter 6 along with Pitman (1938), Welch (1937), and Kendall (1942a, 1945). Wood (1970) gave a transformation that stabilises the variance of the coefficient of concordance.

Weier and Basu (1978) provided a trivariate generalisation of Spearman's rho, and Simon (1977a) generalised Kendall's tau; both procedures are designed for the null hypothesis of total independence between the variables. Patel (1975) extended the Friedman test to right-censored data.

Friedman and Rafsky (1983) extended the generalised correlation coefficient to provide a distribution-free measure for multivariate observations that is sensitive to alternatives involving non-monotonic relationships. Gerig (1969, 1975) and Jensen (1974) extended the Friedman test to the multivariate case. See Puri and Sen (1971).

Henery (1986) showed how to use order statistics models to find average tau between the judges' rankings and the true ranking; this gives a result for tau which is analogous to the result given in Section 7.10 for rho.

Chapter 8

Partial rank correlation

8.1 In interpreting an observed dependence between two qualities we are constantly faced with the question whether an association or correlation of X with Y is really due to the associations or correlations of each with a third quality Z. In the theory of statistics this kind of problem leads to the theories of *partial* association or correlation which attempt to decide the matter by the consideration of sub-populations in which the variation of Z is eliminated. The same problem arises in rank correlation. For instance, if significant correlation appears between mathematical and musical abilities in a number of subjects, the question arises whether this may be attributable to the correlation of each with some more fundamental quality such as intelligence. We proceed here to consider a method of rank correlation which may be applied to an investigation of this kind of problem.

8.2 Suppose we have three rankings of 6 as follows:

$$Z: \quad 1 \quad 2 \quad 3 \quad 4 \quad 5 \quad 6$$
$$X: \quad 3 \quad 1 \quad 4 \quad 2 \quad 6 \quad 5 \,. \qquad\qquad (8.1)$$
$$Y: \quad 4 \quad 2 \quad 1 \quad 6 \quad 3 \quad 5$$

There are $\binom{6}{2} = 15$ possible pairs of individual ranks in each set of rankings.

We take some one set as standard; it does not matter which one we choose. We will take the Z set for convenience, since its ranks are in the natural order. Now we will write down all possible pairs of individual Z ranks and use a + to denote that the corresponding observed pair has the same order as the Z pair and a − to denote that the corresponding observed pair has the opposite order as the Z pair, for each of X and Y. To clarify this procedure, take the Z pair 1, 2. The corresponding X pair is 3, 1 which scores a − ; the corresponding Y pair is 4, 2, which also scores a − . We obtain the results shown in Table 8.1.

For X and Z, the number of agreements is 11 and the number of disagreements is 4 so that $S_{XZ} = 7$. For Y and Z the number of agreements is 9 and the number of disagreements is 6 so that $S_{YZ} = 3$. For X and Y the numbers are 7 and 8 to give $S_{XY} = -1$. The pairwise tau coefficients are then

$$t_{XZ} = \tfrac{7}{15}, \qquad t_{YZ} = \tfrac{3}{15}, \qquad t_{XY} = -\tfrac{1}{15}.$$

Now we look at Table 8.1 and count the number of cases where X and Y both agree with Z (a + in each column); the number where X and Y both

Table 8.1

Z pair	Z	X	Y	Z pair	Z	X	Y
1, 2	+	−	−	2, 6	+	+	+
1, 3	+	+	−	3, 4	+	−	+
1, 4	+	−	+	3, 5	+	+	+
1, 5	+	+	−	3, 6	+	+	+
1, 6	+	+	+	4, 5	+	+	−
2, 3	+	+	−	4, 6	+	+	−
2, 4	+	+	+	5, 6	+	−	+
2, 5	+	+	+				

disagree with Z (a − in each of the X and Y columns); the number where X agrees with Z and Y disagrees with Z (a + in the X column and a − in the Y column), and the number where X disagrees with Z and Y agrees with Z (a − in the X column and a + in the Y column). These counts are presented as a fourfold table in Table 8.2, which sets out the agreements of rankings X and Y with Z.

Table 8.2

		Ranking Y		
		Pairs + (agreeing with Z)	Pairs − disagreeing with Z)	Totals
Ranking X	Pairs + (agreeing with Z)	6	5	11
	Pairs − (disagreeing with Z)	3	1	4
	Totals	9	6	15

Here, for example, there are 11 cases in which X agrees with Z; in 6 of these Y also agrees with Z, and in the remaining 5 it disagrees.

Generally in three rankings of n we shall have the results in Table 8.3.

Table 8.3

a	b	$a + b$
c	d	$c + d$
$a + c$	$b + d$	$N = \binom{n}{2} = a + b + c + d$

We now define a partial rank correlation coefficient of X and Y with Z as

$$t_{XY.Z} = \frac{ad - bc}{\sqrt{[(a + b)(c + d)(a + c)(b + d)]}}. \qquad (8.2)$$

In our present example this becomes

$$\frac{6 - 15}{\sqrt{[(11)(4)(9)(6)]}} = -0.185$$

as compared with $t_{XY} = -0.067$.

8.3   The coefficient of (8.2) is a coefficient of association in a 2×2 table and we have already seen it in another connection in Section 3.14. It can vary from -1 to $+1$ but not outside those limits, and measures the intensity of association between the agreements of X with Z and those of Y with Z.

If the coefficient is unity we have

$$(ad - bc)^2 = (a + b)(a + c)(b + d)(c + d) \tag{8.3}$$

giving

$$4abcd + a^2(bc + bd + cd) + b^2(ac + ad + cd)$$
$$+ c^2(ab + ad + bd) + d^2(ac + ab + bc) = 0. \tag{8.4}$$

Since no a, b, c, d can be negative this can only be true if at least two of them are zero. If two in the same row and column are zero we get the purely nugatory case in which either X or Y is in perfect agreement or disagreement with Z. We have then to consider only $a = 0$ and $d = 0$ or $b = 0$ and $c = 0$. In the latter case X and Y agree completely upon their concordances with Z and $t_{XY.Z} = 1$. In the former case they disagree completely and the coefficient is -1.

8.4   The reader who is acquainted with the chi-square test statistic χ^2 for independence in 2×2 contingency tables can verify that

$$t_{XY.Z} = \sqrt{\frac{\chi^2}{N}}. \tag{8.5}$$

Both statistics therefore measure the degree of departure from the case when the dichotomy of preferences according to X is independent of those according to Y. Suppose, in fact, that they are independent. Then the frequencies in Table 8.2 will be

$$\frac{(a + b)(a + c)}{N} \qquad \frac{(a + b)(b + d)}{N}$$

$$\frac{(a + c)(c + d)}{N} \qquad \frac{(c + d)(b + d)}{N}.$$

The differences between the observed values and these values expected under independence will then be typified by

$$\frac{(a + b)(a + c)}{N} - a = \frac{(a + b)(a + c) - a(a + b + c + d)}{N}$$

$$= \frac{bc - ad}{N}.$$

Thus χ^2 is the sum of four terms like

$$\frac{(bc - ad)^2}{N^2} \bigg/ \frac{(a + b)(a + c)}{N}$$

and the sum reduces to

$$\frac{(bc - ad)^2(a + b + c + d)}{(a + b)(a + c)(c + d)(b + d)}$$

from which (8.5) follows.

8.5　We have therefore constructed a coefficient, capable of varying from -1 to $+1$, which measures the extent to which X and Y agree *so far as concerns their agreement with Z*. If the coefficient is $+1$ they are in complete agreement; if it is zero $ad - bc = 0$ and $a/b = c/d$, so that the preferences are independent; if the coefficient is -1 they are in complete disagreement.

　　We may then say that partial tau as so defined measures the agreement between X and Y independently of the influence of Z. Partial tau is increased by an agreement between X and Y whether they agree with Z or not. The point may be clearer from a further examination of Table 8.3. For the ordinary rank correlation between X and Y we shall have

$$t_{XY} = \frac{(a + d) - (b + c)}{N}. \tag{8.6}$$

In the table, however, we itemise the agreements between X and Y according to whether they do or do not agree with Z. The row containing a, b shows us how far Y has $+$ or $-$ scores in those items for which X has only $+$ scores. If this row is similar (in the sense of proportionality) to the row containing c, d, then X has $+$ or $-$ scores in much the same proportion whether Y has $+$ scores or not. In such circumstances we can hardly regard X and Y as much in agreement, except in so far as they both agree with Z. Our coefficient of partial tau measures the departure from this situation towards greater differences of the X scores whether Y is $+$ or $-$, i.e. gives a better indication that X and Y are more or less in agreement between themselves, whatever the position in regard to Z.

8.6　In addition to (8.6) we have

$$t_{XZ} = \frac{(a + b) - (c + d)}{N} \tag{8.7}$$

$$t_{YZ} = \frac{(a + c) - (b + d)}{N}. \tag{8.8}$$

Remembering that $N = a + b + c + d$, we have

$$1 - t_{XZ}^2 = \frac{4}{N^2}(a + b)(c + d)$$

$$1 - t_{YZ}^2 = \frac{4}{N^2}(a + c)(b + d)$$

$$t_{XY} - t_{XZ}t_{YZ} = \frac{1}{N^2}[(a + b + c + d)\{\overline{a + d} - \overline{b + c}\}$$

$$- \{\overline{a + b} - \overline{c + d}\}\{\overline{a + c} - \overline{b + d}\}]$$

$$= \frac{4}{N^2}(ad - bc)$$

Thus from (8.2),

$$t_{XY.Z} = \frac{t_{XY} - t_{XZ}t_{YZ}}{\sqrt{(1 - t_{XZ}^2)(1 - t_{YZ}^2)}} \tag{8.9}$$

This expresses partial tau in terms of the coefficients tau between the original rankings. It is interesting that this relationship is formally the same as that expressing a partial product-moment correlation in terms of the constituent correlations.

Example 8.1

Three rankings are given according to (1) intelligence, (2) mathematical ability, (3) musical ability. They are as follows:

(1)	1	2	3	4	5	6	7	8	9	10
(2)	1	4	5	6	2	7	3	9	8	10
(3)	4	1	3	5	2	6	7	10	9	8

We find

$$t_{12} = 0.644, \qquad t_{13} = 0.644, \qquad t_{23} = 0.556.$$

Thus, from (8.9),

$$t_{23.1} = \frac{0.556 - (0.644)^2}{1 - (0.644)^2}$$

$$= 0.24.$$

This correlation is weaker than between (2) and (3) without considering (1), and we suspect that correlations between (1) and (2), (1) and (3) may be masking the real relationship between (2) and (3). This kind of inference, however, must be made with considerable reserve. It is a suggestion for further inquiry, nothing more unless there are prior grounds for expecting the effect.

8.7 Tests of significance for the partial tau coefficient can be carried out using Appendix Table 11. This table gives the significance points of $t_{XY.Z}$ for one-tailed tests at levels 0.005, 0.01, 0.025, 0.05. These same significance points apply for two-tailed tests at levels 0.01, 0.02, 0.05, and 0.10.

In the example with the rankings in (8.1) where $n = 6$, our observed value $t_{XY.Z} = -0.185$ is smaller in absolute value that the 0.600 entry in Appendix Table 11 for the one-tailed level 0.05. This means that the probability of obtaining a value as small or smaller than the observed value is greater than 0.05, i.e. *P*-value > 0.05, and thus not significant for one-tailed alternatives $\tau_{XY.Z} < 0$ at any level 0.05 or smaller.

In Example 8.1 where $n = 10$, our observed value 0.24 is again smaller than the 0.413 value for the one-tailed level 0.05 in Appendix Table 11 and hence not significant for one-tailed alternatives $\tau_{23.1} > 0$ at most reasonable levels. When the effect of intelligence is ignored we found $t_{23} = 0.556$ which corresponds to $S = 25$. Appendix Table 1 shows that the corresponding *P*-value is 0.014. Thus our further inquiry has shown that the relations between (1) and (2), (1) and (3), did serve to mask the real relationship between (2) and (3), as we suspected earlier.

8.8 Maghsoodloo (1975) used the following example with $n = 7$ rankings on three variables, arranged so that the *Z* variable follows the natural order (Table 8.4).

Table 8.4

Variable	Subject						
	B	D	C	A	E	G	F
Z	1	2	3	4	5	6	7
X	6	7	5	3	4	1	2
Y	7	6	5	3	4	2	1

For the variables X and Z, $P = 3$, $Q = 18$, $S = -15$, $t_{XZ} = -0.7143$. Also, $t_{YZ} = -0.9048$ and $t_{XY} = 0.8095$ since $S = -19$ for Y and Z and $S = 17$ for X and Y. From Appendix Table 1 the one-tailed *P*-value for X and Y is 0.0054, which is highly significant.

What happens to the relationship between X and Y when we hold variable Z fixed? The calculation from (8.9) is

$$t_{XY.Z} = \frac{0.8095 - 0.6463}{\sqrt{(1 - 0.5102)(1 - 0.8187)}} = 0.548$$

From Appendix Table 11 the one-tailed *P*-value is between 0.025 and 0.05, which is much less significant.

Notes and references

The primary reference for partial tau is Kendall (1942b). A generalised partial correlation coefficient that can be extended to more than three variable was discussed by Somers (1959). A special case is the index of matched correlation discussed by Quade (1974).

Tables of the exact null sampling distribution of partial tau for three variables were developed by Moran (1951) for $n = 3$ and 4. Hoflund (1963) used Monte Carlo simulations to estimate the distribution for rankings of 4 through 10 objects. Maghsoodloo (1975) mentioned a FORTRAN program that gives the exact distributions for any n but the computer time required is large even for $n = 7$. He therefore used Monte Carlo methods to obtain reliable estimates that give tables for n up to 30. These tables were extended to even larger sample sizes by Maghsoodloo and Pallos (1981). This same research concluded that the normal approximation is satisfactory for $n \geq 50$ and gave estimates of the required null variance. The entries in Appendix Table 9 were obtained from these latter two papers.

Hoeffding (1948a) found that the distribution of $\sqrt{n(t_{XY.Z} - \rho_{XY.Z})}$ (where $\rho_{XY.Z}$ is the population partial correlation) is asymptotically normal with zero mean and a variance which has a complicated expression. Shirahata (1980) refined these results. Johnson (1979) investigated the properties of the non-null distribution of $t_{XY.Z}$ and gave formulae for the approximate variance. Henze (1979) obtained the non-null distribution of partial, multiple and tri-correlation rank coefficients.

An early paper on tests based on partial tau is Goodman (1959). Goodman and Grunfeld (1961) applied these tests to time-series data. Other tests of conditional independence were developed by Simon (1977a) and Shirahata (1977).

Another measure of partial correlation can be defined by replacing each t-coefficient for two variables in (8.9) by the corresponding r_s coefficient; this coefficient is a Spearman partial correlation or a partial rho. Shirahata (1980) showed that both the partial rho and tau have asymptotic normal distributions. Davis (1967) proposed a partial coefficient for the Goodman–Kruskal coefficient.

Wolfe (1977) proposed dealing with three variables by calculating the ordinary two-variable tau coefficient between X and the difference $(Z - Y)$ as a measure of the correlation between X and Y and that between Y and Z. Simon (1977a) gave a nonparametric test for total independence of k variables based on the $k(k - 1)/2$ pairwise Kendall taus.

Moran (1951) proposed a nonparametric coefficient of multiple correlation that is analogous to the ordinary theory. For example, with $k = 3$ we have

$$1 - R^2_{1(23)} = (1 - t^2_{13})(1 - t^2_{12.3}).$$

The distribution theory is difficult even for small n. This coefficient is given a new interpretation in Bobko (1977). Lehmann (1977) gave a general derivation of partial and multiple rank correlation.

Partial correlation coefficients computed for contingency tables are discussed in Somers (1959, 1968, 1974), Davis (1967), Simon (1977a, 1977b), and Agresti (1977).

Kendall's tau between two variables, in the presence of a discrete blocking variable, was considered by Quade (1974), Reynolds (1974) and Korn (1984).

Problems

8.1 One possibility for the significant positive relationship observed between students' perceived evaluation of the extent of time teachers addressed science content and context in Problem 4.6 is that both of these perceptions are related to a third variable, such as perhaps students' interest and enthusiasm for science. Compute the partial tau coefficient and determine whether the relationship is explained by students' interest for the data in Table 8.5.

Table 8.5

Class	Content	Context	Interest
1	4.14	3.33	2.14
2	3.83	2.79	2.87
3	3.81	2.61	3.06
4	4.51	3.29	2.62
5	3.65	2.64	3.02
6	4.06	3.10	2.82
7	3.67	2.82	2.91
8	4.42	2.89	2.26
9	3.78	2.81	2.99
10	4.15	3.02	2.39
11	3.82	2.74	2.95

Answer: $t_{12.3} = 0.4365$, $0.025 < P\text{-value} < 0.05$.

8.2 Charlop and Carlson (1983) studied reversal and nonreversal shifts in autistic children aged 2–14 years. Their mental ages ranged from 1.1 to 7.1 as measured by standardised tests. A reversal shift involves teaching a child to respond to one stimulus of a stimulus pair and subsequently reversing the correct answer. Research indicates that ability to make reversal shifts is related to chronological age for normal older children. The study reported usable data on nine autistic older children (Table 8.6). Determine the partial correlation between chronological age (CA) and ability to make reversal shifts (RS) as measured by number of trials to learn the reversal shift, when mental age (MA) is held constant.

Answer: $t = -0.075$

Table 8.6

Child	CA	MA	RS
1	13.3	6.0	10
2	6.5	3.7	72
3	11.2	2.1	27
4	13.1	1.1	29
5	10.2	5.9	10
6	11.1	6.5	11
7	11.6	4.3	15
8	9.3	4.9	25
9	9.7	7.1	10

8.3 The data in Table 8.7 represent the weight, height and age of a random sample of eight nutritionally deficient children in Ethiopia. Find the partial correlation between height and weight when age is held constant.

Table 8.7

Child	Weight	Height	Age
1	49	40	5
2	71	59	10
3	67	62	11
4	55	51	8
5	58	50	7
6	57	48	9
7	51	42	6
8	76	61	12

8.4 In the situation of Problem 1.2, suppose we also have ranks of the ten cities with respect to size (Table 8.8). Determine whether the relationship between pollution and disease (1 = worst) remains significant when size of city (1 = largest) is taken into account.

Table 8.8

City	Pollution	Disease	Size
A	4	5	3
B	7	4	7
C	9	7	9
D	1	3	2
E	2	1	1
F	10	10	10
G	3	2	4
H	5	8	6
I	6	6	5
J	8	9	8

8.5 Dickson, Lusch and Wilkie (1983) reported an experiment to replicate the Guttman scaling studies which suggest there is an underlying common order of acquisition of home appliances. The study used Kendall's tau and found a significant positive relationship between the Guttman scale and average years of ownership (a surrogate measure of order of acquisition) for the data shown in Table 8.9 and based on a sample of 3311 households.

Suppose that information had also been obtained on average current income of these 3311 households with the results shown in the last column. Determine whether the significant positive relationship remains when the effect of income is removed.

Table 8.9

Appliance	Guttman scale	Average years of ownership	Average income (000)
Refrigerator	1	15.4	21.3
Clothes washer	2	12.5	19.4
Colour TV	3	6.0	24.2
Sewing machine	4	12.0	10.3
Range oven	5	11.3	16.8
Clothes dryer	6	8.4	17.2
Stereo AM/FM radio	7	5.2	28.3
Separate freezer	8	5.5	30.6
Dishwasher	9	3.4	23.9
Room air-conditioner	10	3.0	28.5
Microwave oven	11	0.3	26.8
Video recorder	12	0.1	32.6

8.6 Scores made by seven boys on a reading test, an arithmetic test, and a standard IQ test are shown in Table 8.10. Compute the Kendall partial correlation coefficient between reading and arithmetic scores when the effect of IQ scores is eliminated and interpret the result.

Table 8.10

Student	X_1 Reading	X_2 Arithmetic	X_3 IQ
A	53	48	105
B	47	55	103
C	56	57	104
D	69	63	118
E	32	41	92
F	55	58	110
G	63	64	113

Answers: $t_{12} = 0.71429$, $t_{13} = 0.80952$, $t_{23} = 0.71429$, $t_{12.3} = 0.33114$, P-value > 0.05.

Chapter 9

Ranks and variate values

9.1 Up to this point we have considered ranks as the fundamental data of a given statistical situation, irrespective of the manner in which they were reached. In many such situations, however, the ranking takes place (or is supposed to take place) according to the values of a statistical variable or *variate*. It is of considerable interest to consider the relationships between the ranks and the corresponding variate values, or between measures of correlation based on ranks and those based on the variate values.

Concordances

9.2 In general we shall consider a continuous population, that is to say a population for which the variate values may be any of a continuous range. One point to notice initially is that elements in such a population cannot possess a rank correlation, strictly speaking; for the essence of ranking is that the objects can be placed in a meaningful order and the totality of values of a continuous variate cannot be ordered in this sense.

9.3 We can nevertheless relate the ideas of correlation of ranks and correlation of variates by considering *order properties*. Suppose we draw two members x_i and x_j at random from a continuous population. The probability that they are equal is zero and thus the possibility that $x_i = x_j$ can be ignored. We may then consider the probability that $x_i < x_j$ and the complementary probability that $x_i > x_j$. Moreover, if we draw two members x_i, y_i and x_j, y_j from a bivariate population we may consider the probabilities of *concordance* of type 1:

$$\pi_1 = \text{Prob}(y_i < y_j \mid x_i < x_j), \tag{9.1}$$

that is to say, the conditional probability that $y_i < y_j$ given that $x_i < x_j$. We have the complementary probability

$$1 - \pi_1 = \text{Prob}(y_i > y_j \mid x_i < x_j). \tag{9.2}$$

The probability π_1 represents a property of the population.

9.4 Now suppose that we draw a sample of n values at random from the bivariate population and arrange the xs in ascending order of magnitude. Of the $\frac{1}{2}n(n-1)$ pairs of xs which we may choose for comparison, some have the corresponding ys in ascending order and some do not. The number of those which do divided by $\frac{1}{2}n(n-1)$ is clearly an estimator of π_1. Moreover it is an unbiased estimator, as we shall prove in the next chapter. If p_1 is this proportion and $q_1 = 1 - p_1$ we see at once that the tau coefficient for this sample, is simply given by

$$t = p_1 - q_1 = 2p_1 - 1. \tag{9.3}$$

We could therefore define t (as we have already noticed in Section 1.12 and 2.14) in terms of concordance and arrive at a coefficient which has an analogue in the continuous case.

9.5 Suppose now that we have three pairs of values (x_i, y_i), (x_j, y_j), (x_k, y_k), $i \neq j \neq k$. We define the probability of concordance of type 2 as

$$\pi_2 = \text{Prob}(y_i < y_k \mid x_i < x_j), \tag{9.4}$$

the probability that y_i is less than y_k, given that $x_i < x_j$. We define a sample quantity p_2 as the number of concordances of type 2 in the sample divided by the total possible number. Unlike p_1, which can vary from 0 to 1, p_2 (as we shall show in the next chapter) can only vary from $\frac{1}{3}$ to $\frac{2}{3}$. For reasons which will appear later we do not take $6(p_2 - \frac{1}{2})$, which can vary from -1 to $+1$, as another coefficient. Instead, following (2.35) we define a sample value

$$r_s = \frac{3t}{n+1} + \frac{6(n-2)}{n+1}(p_2 - \tfrac{1}{2}). \tag{9.5}$$

The coefficient r_s is Spearman's coefficient for the sample, and again it appears that we may define a rank correlation coefficient in terms of concordances. We note that for large n (9.5) becomes

$$\rho_s = 6(\pi_2 - \tfrac{1}{2}), \tag{9.6}$$

which may be regarded as a definition of Spearman's rho for a continuous population.

Relation between ranks and variate values

9.6 Suppose that we draw a sample from a univariate population of x-values and rank them in ascending order of magnitude. It is of some interest to consider the product-moment coefficient of correlation between the x-values and the ranks. The coefficient is, in fact, sometimes surprisingly high. The basic results (due to A. Stuart) are as follows:

If the correlation for sets of n is C_n, and the limiting value as n tends to infinity is C, we always have

$$C_n = \left(\frac{n-1}{n+1}\right)^{1/2} C. \tag{9.7}$$

If the parent population is a continuous uniform distribution, then $C = 1$ and

$$C_n = \left(\frac{n-1}{n+1}\right)^{1/2}. \tag{9.8}$$

If the parent population is a normal distribution, then $C = \sqrt{(3/\pi)}$ and

$$C_n = 0.9772\left(\frac{n-1}{n+1}\right)^{1/2}. \tag{9.9}$$

If the parent population is a Gamma distribution

$$dF = \frac{1}{\Gamma(m)} e^{-x} x^{m-1} \, dx, \qquad 0 < x < \infty$$

then

$$C_n = \left(\frac{n-1}{n+1}\right)^{1/2} \left(\frac{3m}{\pi}\right)^{1/2} \frac{\Gamma(m+\frac{1}{2})}{\Gamma(m+1)}. \tag{9.10}$$

For example, in rankings of 10 from a normal population the product-moment correlation between ranks and variates is

$$0.9772\sqrt{\tfrac{9}{11}} = 0.884.$$

9.7 In view of this fairly close relationship between ranks and variates we might expect that if we replace variate values by rank numbers and then operate on the latter as if they were the primary variates we should in many cases draw the same conclusions. This appears to be so in a number of practical cases; but the procedure has to be followed with a certain amount of caution. The replacement with ranks has effectively standardised the scale of the variate and fixed the mean, a procedure which might in some instances lead us astray.

9.8 A converse procedure has been recommended by some authors. Given a ranking of n, we replace the ranks by variate values x; the value x_i is the expected value of the ith order statistic of a sample of n drawn from a normal population. This, of course, does not obviate the difficulty referred to in Section 9.7. It has, however, been shown by Hoeffding (1951) that when tests concerning normal hypotheses are under examination such a procedure has optimum properties. The replacement of variates by ranks, from the geometrical viewpoint, is equivalent to approximating the sample distribution function by a straight line; the replacement by equivalent normal deviates amounts to approximating by a normal distribution.

Relation between t and parent correlation in the normal case

9.9 We now need to modify and to extend our notation slightly to avoid confusion.

(1) The parameter of the bivariate normal population expressing the product-moment correlation will be denoted by ρ. The product-moment correlation of a sample of variates will be denoted by r.

(2) We shall sometimes make estimates of ρ from values of t or r_s. These we shall denote by primes on r, e.g. r'.

9.10 It may be shown that for samples from a normal population

$$E(t) = \frac{2}{\pi} \sin^{-1}\rho. \tag{9.11}$$

For instance, if $\rho = 1/\sqrt{2} = 0.707$, $E(t) = 0.5$. We may therefore construct an estimator of ρ, say r', by putting

$$r' = \sin \tfrac{1}{2}\pi t. \tag{9.12}$$

This is *not* an unbiased estimator of ρ, for we should require that $E(r') = \rho$, which does not follow from (9.11). Nevertheless, the procedure appears reasonable. The relation (9.11) was given for the first time by Greiner (1909) and the expression (9.13) below for its variance by Esscher (1924).

9.11 In the next chapter we shall show that for normal samples

$$\text{var } t = \frac{2}{n(n-1)} \left\{ 1 - \left(\frac{2}{\pi} \sin^{-1}\rho \right)^2 + 2(n-2) \left[\tfrac{1}{9} - \left(\frac{2}{\pi} \sin^{-1}\tfrac{1}{2}\rho \right)^2 \right] \right\}. \tag{9.13}$$

We know that t is normally distributed for large samples and hence may use this result to test the significance of an observed t. But to do so we have to assume some values for the unknown ρ. In accordance with the usual practice in the theory of large samples we shall replace ρ by r' of equation (9.12).

If p and q are the proportion of positive and negative scores contributing to t we have

$$t = p - q$$

and hence

$$1 - t^2 = 4pq. \tag{9.14}$$

Thus from (9.13) we have, for large samples

$$\text{var } t = \frac{8}{n(n-1)} \left\{ pq + \tfrac{1}{2}(n-2) \left[\tfrac{1}{9} - \left(\frac{2}{\pi} \sin^{-1}\tfrac{1}{2}r' \right)^2 \right] \right\}. \tag{9.15}$$

It may also be shown that

$$0 \le \tfrac{1}{9} - \left(\frac{2}{\pi} \sin^{-1}\tfrac{1}{2}r' \right)^2 \le \tfrac{4}{9}pq \tag{9.16}$$

and, for $n \ge 10$

$$\frac{2}{n(n-1)} [1 + \tfrac{2}{9}(n-2)] \le \frac{5}{9(n-1)}. \tag{9.17}$$

Substituting in (9.15) we find

$$\text{var } t \le \frac{20pq}{9(n-1)} = \frac{5(1-t^2)}{9(n-1)}. \tag{9.18}$$

The upper limit is, in fact, attained in a number of cases, so that the equality in (9.18) gives us a fairly good estimate of the actual variance.

9.12 We compare the result in (9.18) with (4.12) written in the form

$$\text{var } t \le \frac{2(1-t^2)}{n}. \tag{9.19}$$

Apart from the difference in the factors n and $n-1$, which is not important for large samples, we see on comparing (9.18) with (9.19) that the former gives a limit which is only 0.278 times that of the latter, the standard error being 0.53 or little more than half as great. This is the gain in accuracy which we acquire at the expense of assuming that the population is normal.

9.13 Since

$$r' = \sin \tfrac{1}{2}\pi t$$

we have, for small variations,

$$dr' = \tfrac{1}{2}\pi \cos(\tfrac{1}{2}\pi t)dt.$$

Squaring and summing for all such variations we find

$$\text{var } r' = \tfrac{1}{4}\pi^2(1 - r'^2) \text{ var } t. \tag{9.20}$$

By the use of (9.18) this gives us

$$\text{var } r' \le \frac{5\pi^2}{9} pq \frac{1-r'^2}{n-1}, \qquad n \ge 10$$

$$\le (2.34)^2 pq \frac{1-r'^2}{n-1}. \tag{9.21}$$

If we use r' to estimate p, (9.16) provides an estimate of the upper limit to the standard error of the estimate. It is interesting to compare this with the standard error of the product-moment sample coefficient r, which is given by

$$\text{var } r = \frac{(1-r^2)^2}{n}. \tag{9.22}$$

Taking the upper limit in (9.21) and ignoring the difference between n and $n-1$ we have

$$\sqrt{\frac{\text{var } r'}{\text{var } r}} = \frac{2.34\sqrt{(pq)}}{\sqrt{(1-r^2)}}. \tag{9.23}$$

If the parent p is zero, p is approximately equal to q, so the ratio of standard errors given by (9.23) is approximately 1.17. If $\rho = 0.9$ we may put, approximately,

$$r = 0.9, \qquad t = \frac{2}{\pi}\sin^{-1}0.9 = 0.713, \qquad pq = \tfrac{1}{4}(1 - t^2) = 0.123.$$

The ratio of standard errors then becomes approximately 1.88. The product-moment coefficient r is more accurate that r' in the sense that it has a smaller standard error and therefore is more likely to be nearer the true value ρ.

Relation between ρ_s and ρ in the normal case

9.14 If we define τ for a continuous population by the expression analogous to (9.3),

$$\tau = 2\pi_1 - 1, \tag{9.24}$$

then t is an unbiased estimator of τ. But proceeding likewise from (9.5) gives us

$$E(r_s) = \frac{3\tau}{n + 1} + \frac{6(n - 2)}{n + 1}(\pi_2 - \tfrac{1}{2}),$$

and if we define ρ_s for the population by (9.6) this gives us

$$E(r_s) = \frac{n - 2}{n + 1}\rho_s + \frac{3\tau}{n + 1}$$

$$= \rho_s + \frac{3(\tau - \rho_s)}{n + 1}. \tag{9.25}$$

This confirms the result of (5.76), that r_s is not an unbiased estimator of ρ_s.

9.15 Let us now consider the relation between ρ_s and ρ in a normal population. It may be shown that

$$\rho = 2\sin\tfrac{1}{6}\pi\rho_s,$$

and hence we may take as an estimator of ρ for large samples the quantity r'' where

$$r'' = 2\sin\tfrac{1}{6}\pi r_s. \tag{9.26}$$

In view of the bias revealed by (9.25), however, it seems better to take

$$r'' = 2\sin\tfrac{1}{6}\pi\left[r_s - \frac{3(t - r_s)}{n - 2}\right]. \tag{9.27}$$

Formula (9.26) is due to Pearson (1907) and was arrived at by considering grade correlations as follows.

9.16 We define the *grade* of an individual variate value x as the proportional frequency of the population with variate values less than or equal to x. The correlation between the grades of variates x, y in a bivariate correlation is called the *grade correlation*. This quantity exists for a continuous population and is easily seen to reduce to the Spearman rank correlation when applied to a finite sample. By considering concordances, we shall prove in the next chapter that the grade correlation as so defined is in fact the quantity ρ_s defined by (9.6).

9.17   It might be expected that there would be some reasonably tractable formula like (9.13) giving the variance of r_s in the normal case. This is not so and, in fact, it has been shown that no such formula can exist in terms of elementary functions.

For large samples we have from (9.26)

$$\frac{dr''}{dr_s} = \tfrac{1}{3}\pi \cos \tfrac{1}{6}\pi r_s$$

and hence

$$\text{var } r'' = \tfrac{1}{9}\pi^2(1 - \tfrac{1}{4}r^2) \text{ var } r_s. \tag{9.28}$$

This is only of help to us when we know var r_s. In the case when $\rho = 0$ this reduces to $1/(n-1)$ and thus

$$\text{var } r'' = \frac{\pi^2}{9(n-1)}$$

$$= \frac{(1.047)^2}{n-1}. \tag{9.29}$$

When ρ is not zero an expression (valid for large n) can be derived in the form of an infinite series as follows:

$$\text{var } r_s = \frac{1}{n}(1 - 1.5635\rho^2 + 0.3047\rho^4 + 0.1553\rho^6$$

$$+ 0.0616\rho^8 + 0.0242\rho^{10} + \cdots). \tag{9.30}$$

Improved versions of such formulae have been given by Fieller, Hartley and Pearson (1957), David and Mallows (1961), and Fieller and Pearson (1961). The latter give some tables and verify by sampling experiments. They also consider the results of a normalising transformation $z_s = \tanh^{-1} r_s$ or $z_t = \tanh^{-1} t$ and suggest that it is sufficient to test these transformed quantities using the normal distribution with

$$\text{var } z_s = \frac{1.060}{n-3}, \quad \text{var } z_t = \frac{0.437}{n-4}.$$

9.18   If we compare the standard error of r'' with that given by (9.22) we find for $\rho = 0$ a ratio of 1.047 and for $\rho = 0.5$ a ratio of 1.137. The balance of precision still lies in favour of product-moment r but not very strongly so, and an estimator of ρ based on r_s, all things considered, is quite good. This bears out the remarks of Section 9.7. It is, in fact, not an accident, though we cannot enter here into the reasons, that the variance of r'' in (9.29) has a factor $\pi^2/9$ whereas the factor C in (9.9) is the reciprocal of the fourth root of this quantity.

9.19 It is also of some interest to consider the relationship between t and r_s for the normal case. For large samples an expansion similar to that of (9.30) gives

$$\text{cov}(r_s, t) = \frac{2}{3n}(1 - 1.2486\rho^2 + 0.0683\rho^4 + 0.0728\rho^6$$

$$+ 0.0403\rho^8 + 0.0164\rho^{10} + \cdots). \tag{9.31}$$

We find, for example, that for $\rho = 0$ the correlation between r_s and t is unity, as we already know from Section 5.14; for $\rho = 0.2$ it is 0.999 55; for $\rho = 0.4$ it is 0.9981; and even for $\rho = 0.8$ it is 0.9843 though it tends to zero as ρ approaches unity. For large n the ratio of r_s and t tends to the ratio of their expectations, in spite of the fact that the correlation between them may be small for high ρ, because their variances tend to zero. This ratio is $3 \sin^{-1}\frac{1}{2}\rho / \sin^{-1}\rho$, which varies from 1.3 in the neighbourhood of $\rho = 0$ through 1.42 when $\rho = 0.6$ to unity when $\rho = 1$. This is one contributory reason why, in practice, we often find a value of r_s about 40 per cent or 50 per cent greater than t, unless either is near unity.

9.20 In conclusion let us remove one possible misunderstanding arising from the relative magnitudes of the standard errors of r' and r'', namely the estimators of ρ based on t and r_s. The fact that they are different indicates a genuine difference of efficiency in the estimation process, the one with the smaller variance being the more efficient.

Now for large samples the variances of t and r_s as given by (9.13) and (9.30) are also different. But this does not mean that t is a better or worse estimator of τ than r_s is of ρ_s. The difference in variances is due to the difference of scales; and it is easy to verify that for ρ not close to 1 we have

$$\frac{E(t)}{\sqrt{\text{var } t}} = \frac{E(r_s)}{\sqrt{\text{var } r_s}}, \tag{9.32}$$

which is in accordance with our general result that the correlation of t and r_s is high, at least for large n and ρ not close to 1.

Notes and references

See Esscher (1924) and Greiner (1909) for the mean and variance of t. For the mean of r_s see Hoeffding (1959a) and Moran (1948) and for the variance see Kendall (1949) and David, Kendall and Stuart (1951). See also references to the next chapter. For grade correlations see Pearson (1907).

For the relationship between ranks and variate values see Stuart (1954a). The asymptotic relation for the normal case had previously been obtained but not published by Sir Cyril Burt.

For concordances of the first and second kind see Sundrum (1953d).

Montjardet and LeConte de Poly-Barbut (1986) studied extremal values of the difference between the two rank correlation coefficients rho and tau.

Chapter 10

Proof of the results of Chapter 9

Correlation between ranks and variate values

10.1 Let N samples of size n be drawn from a continuous population with mean μ and variance σ^2. In each sample the observations are ordered, the ith smallest value being given the rank i. Let us evaluate, for the set of Nn observations, the covariance of the variate values and ranks (μ_{11}) and the variances of variate values (μ_{20}) and ranks (μ_{02}).

If the ith smallest value in the jth sample is $x_{(i)j}$ we have

$$\mu_{02} = \tfrac{1}{12}(n^2 - 1) \tag{10.1}$$

$$\mu_{20} = \frac{1}{Nn} \sum_{i=1}^{n} \sum_{j=1}^{N} \left\{ x_{(i)j} - \frac{1}{Nn} \sum_{i=1}^{n} \sum_{j=1}^{N} x_{(i)j} \right\}^2$$

$$= \frac{1}{Nn} \sum_{i} \sum_{j} x_{(i)j}^2 - \left\{ \frac{1}{Nn} \sum_{i} \sum_{j} x_{(i)j} \right\}^2 \tag{10.2}$$

$$\mu_{11} = \frac{1}{Nn} \sum_{i} \sum_{j} [i - \tfrac{1}{2}(n + 1)] \left(x_{(i)j} - \frac{1}{Nn} \sum_{i} \sum_{j} x_{(i)j} \right). \tag{10.3}$$

Now as N tends to infinity

$$\lim \frac{1}{N} \sum_{j} x_{(i)j}^r = E(x_{(i)}^r)$$

and hence

$$\lim_{N \to \infty} \mu_{20} = \frac{1}{n} \sum_{i} E(x_{(i)}^2) - \left[\frac{1}{n} \sum_{i} E(x_{(i)}) \right]^2, \tag{10.4}$$

$$\lim_{N \to \infty} \mu_{11} = \frac{1}{n} \sum_{i} iE(x_{(i)}) - \left\{ \frac{n + 1}{2n} \sum_{i} E(x_{(i)}) \right\}^2. \tag{10.5}$$

The distribution of $x_{(i)}$, say $G(x)$, in samples of n from a population with distribution function $F(x)$ is given by

$$dG(x) = n \binom{n - 1}{i - 1} [F(x)]^{i-1} [1 - F(x)]^{n-i} \, dF(x). \tag{10.6}$$

Thus

$$\frac{1}{n} \sum_{i} E(x_{(i)}^r) = \int_{-\infty}^{\infty} \left[\sum_{i} \binom{n - 1}{i - 1} F^{i-1}(1 - F)^{n-i} \right] x^r \, dF(x) \tag{10.7}$$

But the sum in brackets is the binomial expansion of $\{F + (1 - F)\}^{n-1}$ and is therefore unity. Hence

$$\frac{1}{n} \sum_i E(x_{(i)}^r) = \int_{-\infty}^{\infty} x^r \, dF$$

$$\left.\begin{array}{ll} = \mu & \text{if } r = 1 \\ = \sigma^2 + \mu^2 & \text{if } r = 2 \end{array}\right\}. \tag{10.8}$$

We shall assume that these moments exist. Similarly we find

$$\frac{1}{n} \sum_i iE\{x_{(i)}\} = (n - 1) \int_{-\infty}^{\infty} xF \, dF + \mu. \tag{10.9}$$

Hence from (10.4) and (10.5) we find

$$\lim_{N \to \infty} \mu_{20} = \sigma^2, \tag{10.10}$$

$$\lim \mu_{11} = (n - 1) \left[\int_{-\infty}^{\infty} xF(x) \, dF(x) - \tfrac{1}{2}\mu \right]. \tag{10.11}$$

Thus the correlation required is given by

$$C_n = \lim_{N \to \infty} \frac{\mu_{11}}{\sqrt{(\mu_{20}\mu_{02})}}$$

$$= \left[\frac{12(n - 1)}{\sigma^2(n + 1)} \right]_{1/2} \left[\int_{-\infty}^{\infty} xF(x) \, dF - \tfrac{1}{2}\mu \right]. \tag{10.12}$$

If we now let n tend to infinity we get

$$C = \lim_{n \to \infty} C_n = \left(\frac{12}{\sigma^2} \right)^{1/2} \left[\int_{-\infty}^{\infty} xF(x) \, dF - \tfrac{1}{2}\mu \right], \tag{10.13}$$

so that

$$C_n = \left(\frac{n - 1}{n + 1} \right)^{1/2} C. \tag{10.14}$$

10.2   We now develop the value of C for some particular cases.
If the expression in brackets in (10.13) is denoted by $\tfrac{1}{4}\Delta$ we have

$$\Delta = 4 \int_{-\infty}^{\infty} x[F(x) - \tfrac{1}{2}] \, dF(x)$$

which, after integration by parts, gives

$$\Delta = 2 \int_{-\infty}^{\infty} F(x)[1 - F(x)] \, dx$$

$$= 2 \int_{-\infty}^{\infty} \int_{-\infty}^{\infty} [|x - y| \, dF(y)] \, dF(x). \tag{10.15}$$

This quantity Δ is, in fact, the coefficient of dispersion known as Gini's mean difference and is non-negative. We then obtain Stuart's formula

$$C = \frac{\Delta\sqrt{3}}{2\sigma}. \tag{10.16}$$

10.3 For the uniform distribution

$$dF = \frac{dx}{k}, \qquad 0 \le x \le k, \tag{10.17}$$

we have

$$\sigma^2 = \tfrac{1}{12}k^2,$$

$$\Delta = \tfrac{1}{3}k,$$

and hence

$$C = 1. \tag{10.18}$$

This is as we should expect for a distribution in which all values are equally probable.

10.4 For the normal distribution, which we may take to have unit variance and zero mean without loss of generality,

$$dF = \frac{1}{\sqrt{(2\pi)}} e^{-(1/2)x^2} dx, \qquad -\infty < x < \infty, \tag{10.19}$$

we have

$$\sigma^2 = 1$$

$$\Delta = \frac{2}{\sqrt{\pi}}$$

and hence

$$C = \sqrt{\frac{3}{\pi}}. \tag{10.20}$$

10.5 For the Gamma distribution

$$dF = \frac{1}{\Gamma(m)} e^{-x} x^{m-1} dx, \qquad 0 < x < \infty, \tag{10.21}$$

we find

$$\sigma^2 = m$$

$$\Delta = \frac{m\Gamma(2m+1)}{2^{2m-1}[\Gamma(m+1)]^2}$$

whence, using the formula

$$\pi^{1/2}\Gamma(2m+1) = 2^{2m}\Gamma(m+\tfrac{1}{2})\Gamma(m+1),$$

we find

$$C = \frac{\Gamma(m + \frac{1}{2})}{\Gamma(m + 1)} \sqrt{\frac{3m}{\pi}}. \tag{10.22}$$

These are the results stated in Section 9.6. For $m = \frac{1}{2}$ (the smallest value of any statistical interest), $C = 0.78$; for $m = 1$, $C = 0.87$; for $m = 4$, $C = 0.95$.

Concordance

10.6　We now show that p_1 and p_2, the sample relative-frequencies of concordance, are unbiased estimators of the parent values π_1 and π_2. From one approach to probability theory this is obvious, but perhaps a simple proof is not out of place.

Let us attach to any pair of members a variate which is unity if there is a concordance of type 1 and zero otherwise. The expectation of p_1 is then the expectation of this variate for any given pair, since the expectation of a sum is the sum of expectations even when the constituent members are dependent. But the expectation of this variate is π_1 and the result follows. The same line of proof applies to p_2 and π_2.

10.7　Let the distribution function of x, y be $F(x, y)$. The distribution functions of x or y alone are respectively $F(x, \infty)$ and $F(\infty, y)$.

Now for any fixed x_j the probability that $x_i < x_j$ is $F(x_j, \infty)$ and hence the probability that $y_i < y_j$ given $x_i < x_j$ is $F(x_j, y_j)/F(x_j, \infty)$. To obtain p_1, the probability that for any two pairs, (x_i, y_i) and (x_j, y_j), $y_i < y_j$ if $x_i < x_j$, we integrate with respect to x_j, y_j to obtain

$$\pi_1 = \frac{\iint F(x_j, y_j)\, dF(x_j, y_j)}{\iint F(x_j, \infty)\, dF(x_j, y_j)}. \tag{10.23}$$

We may now drop the suffix j. Moreover

$$\iint F(x, \infty)\, dF(x, y) = \int F(x, \infty)\, dF(x, \infty)$$

$$= [\tfrac{1}{2}F^2(x, \infty)]_0^1 = \tfrac{1}{2}$$

and hence

$$\pi_1 = 2 \iint F(x, y)\, dF(x, y), \tag{10.24}$$

with the integration over the whole range of x and y.

Likewise we find, by a similar type of argument,

$$\pi_2 = 2 \iint F(x, \infty)F(\infty, y)\, dF(x, y). \tag{10.25}$$

10.8　Consider the case when the variates are perfectly related by a linear relation. The joint distribution of x and y then becomes univariate and without loss of generality we may suppose this distribution to be uniform in the range 0 to 1. $F(x, y)$ is then reducable to a single variate z, say, and

$$\pi_1 = 2 \int_0^1 z \, dz = 1.$$

In the case of independence $\pi_1 = 0$ and thus π_1 may vary from 0 to 1. It cannot lie outside those limits because it is a probability.

But for π_2 we have, in the case of complete linear dependence,

$$\pi_2 = 2 \int_0^1 z^2 \, dz = \tfrac{2}{3},$$

and if there is negative dependence

$$\pi_2 = 2 \int_0^1 z(1 - z) \, dz = \tfrac{1}{3}.$$

These are extreme values and it is easy to show (cf. Section 2.8) that π_2 cannot lie outside the range $\tfrac{1}{3}$ to $\tfrac{2}{3}$. The coefficient $6(\pi_2 - \tfrac{1}{2})$ accordingly lies between -1 and 1.

10.9 From (10.25) we see that the covariance of the grades of x and y is

$$\text{cov} = \int \int F(x, \infty) F(\infty, y) \, dF(x, y)$$

$$- \left[\int F(x, \infty) \, dF(x, \infty) \right] \left[\int F(\infty, y) \, dF(\infty, y) \right]$$

$$= \tfrac{1}{2}\pi_2 - \tfrac{1}{4}.$$

The variance becomes $\tfrac{1}{3} - \tfrac{1}{4} = \tfrac{1}{12}$. Hence the grade correlation is

$$\tfrac{1}{2}(\pi_2 - \tfrac{1}{2})/\tfrac{1}{12} = 6(\pi_2 - \tfrac{1}{2}), \tag{10.26}$$

and is thus the quantity we have defined as ρ_s.

10.10 Now we proceed to the derivation of the expressions for means and variance of t and r_s in the normal case.

Let sgn ξ stand for $+1$ if ξ is positive, zero if ξ is zero and -1 if ξ is negative. We shall require the result that for real ξ

$$\text{sgn } \xi = \frac{1}{\pi} \int_{-\infty}^{\infty} \frac{e^{it\xi} \, dt}{it}$$

$$= +1, \qquad \xi > 0 \tag{10.27}$$

$$= 0, \qquad \xi = 0$$

$$= -1, \qquad \xi < 0.$$

The integral is to be understood as a principal value, that is,

$$\int_{-\infty}^{\infty} = \lim_{\substack{c \to \infty \\ \varepsilon \to 0}} \left(\int_{-c}^{-\varepsilon} + \int_{\varepsilon}^{c} \right).$$

Equation (10.27) is equivalent to the real integral

$$\operatorname{sgn} \xi = \frac{1}{\pi} \int_{-\infty}^{\infty} \frac{\sin t\xi \, dt}{t}. \tag{10.28}$$

From the definition as a principal value it is clear that if $\xi = 0$ the integral vanishes because of the symmetry of the integrand. Perhaps the quickest way of establishing (10.27) is to consider the complex integral

$$\int \frac{e^{iz\xi} \, dz}{iz}.$$

If ξ is positive we take this integral around the contour consisting of the real axis from $-R$ to $-\varepsilon$, the small semicircle of radius ε above the axis, the real axis from ε to R and the large semicircle of radius R above the axis. This integral vanishes, for the integrand has no poles inside the contour. The integral along the real axis tend to

$$\int_{-\infty}^{\infty} \frac{e^{it\xi} \, dt}{it}.$$

The integral round the large semicircle tends to zero as R tends to infinity. The integral round the small semicircle is effectively the integral of dz/iz round that semicircle clockwise and is $-\pi$. Thus

$$\int_{-\infty}^{\infty} \frac{e^{it\xi} \, dt}{it} - \pi = 0,$$

whence the result (10.27) for $\xi > 0$ follows. If $\xi < 0$ we consider the integral with the sign of t changed.

10.11 Consider now a bivariate normal population of variates x, y with correlation ρ and distribution

$$dF = \frac{1}{2\pi\sqrt{(1 - \rho^2)}} \exp\left[-\frac{1}{2(1 - \rho^2)} (x^2 - 2\rho xy + y^2) \right] dx \, dy. \tag{10.29}$$

We lose no generality by taking the variates measured from zero means with unit variances. If we take a pair of values of x, say x_1 and x_2. we may allot a score based on $x_1 - x_2$, and for the calculation of τ can take this score to be $\operatorname{sgn}(x_1 - x_2)$ or some convenient positive numerical multiple of $x_1 - x_2$. The distribution of a pair of independent values x_1 and x_2, y_1 and y_2 is

$$dF \propto \exp\left\{ -\frac{1}{2(1 - \rho^2)} [x_1^2 + x_2^2 - 2\rho(x_1 y_1 + x_2 y_2) + y_1^2 + y_2^2] \right\} dx_1 \, dx_2 \, dy_1 \, dy_2.$$

$$\tag{10.30}$$

Make the transformation

$$u_1 = \frac{1}{\sqrt{2}}(x_1 - x_2), \qquad u_2 = \frac{1}{\sqrt{2}}(x_1 + x_2)$$

$$v_1 = \frac{1}{\sqrt{2}}(y_1 - y_2), \qquad v_2 = \frac{1}{\sqrt{2}}(y_1 + y_2).$$

The distribution then becomes

$$dF \propto \exp\left[-\frac{1}{2(1 - \rho^2)}(u_1^2 - 2\rho u_1 v_1 + v_1^2)\right] du_1\, dv_1$$

$$\times \exp\left[-\frac{1}{2(1 - \rho^2)}(u_2^2 - 2\rho u_2 v_2 + v_2^2)\right] du_2\, dv_2. \qquad (10.31)$$

Consequently u_1 and v_1 are also distributed normally with correlation ρ independently of u_2 and v_2. Dropping the suffixes we have

$$dF \propto \exp\left[-\frac{1}{2(1 - \rho^2)}(u^2 - 2\rho uv + v^2)\right] du\, dv. \qquad (10.32)$$

10.12 Now if t is a sample value of τ, $E(t)$ is the expectation of the sum of $\frac{1}{2}n(n - 1)$ terms each of which may be written sgn u sgn v. Thus

$$E(t) = E(\text{sgn } u \text{ sgn } v)$$

$$= \int_{-\infty}^{\infty} \int_{-\infty}^{\infty} \text{sgn } u \text{ sgn } v\, dF,$$

which because of (10.27) becomes

$$\frac{1}{\pi^2} \int_{-\infty}^{\infty} \frac{dt_1}{it_1} \int_{-\infty}^{\infty} \frac{dt_2}{it_2} \left(\int_{-\infty}^{\infty} \int_{-\infty}^{\infty} e^{iut_1 + ivt_2}\, dF\right). \qquad (10.33)$$

The expression in parenthesis is the characteristic function of u and v and is equal to

$$\exp -\tfrac{1}{2}(t_1^2 + 2\rho t_1 t_2 + t_2^2). \qquad (10.34)$$

Hence

$$E(t) = \frac{1}{\pi^2} \int_{-\infty}^{\infty} \int_{-\infty}^{\infty} \frac{dt_1\, dt_2}{it_1\, it_2} \exp -\tfrac{1}{2}(t_1^2 + 2\rho t_1 t_2 + t_2^2)$$

Thus

$$\frac{\partial E(t)}{\partial \rho} = \frac{1}{\pi^2} \int_{-\infty}^{\infty} \int_{-\infty}^{\infty} dt_1\, dt_2 \exp -\tfrac{1}{2}(t_1^2 + 2\rho t_1 t_2 + t_2^2)$$

$$= \frac{2}{\pi\sqrt{(1 - \rho^2)}}. \qquad (10.35)$$

Hence, by a simple integration for ρ, remembering that $E(t)$ vanishes when

$\rho = 0$, we have

$$E(t) = \frac{2}{\pi} \sin^{-1} \rho, \tag{10.36}$$

the result given in (9.11).

10.13   To find the variance of t in all possible samples we require

$$E(t^2) = E(\sum \text{sgn } u \text{ sgn } v)^2$$

where summation extends over all $\binom{n}{2}$ values of u and v. We may write

$$\sum(\text{sgn } u \text{ sgn } v)^2 = \sum \text{sgn } u_{ij} \text{ sgn } u_{kl} \text{ sgn } v_{ij} \text{ sgn } v_{kl}$$

and there are three cases:
   (i) If $i = k, j = l$ the term is $+1$ and the expectation of each term is $+1$.
   (ii) If $i \neq k, j \neq l$, the expectation of the term reduces to

$$E(\text{sgn } u_{ij} \text{ sgn } v_{ij})E(\text{sgn } u_{kl} \text{ sgn } v_{kl}) = \left(\frac{2}{\pi} \sin^{-1} \rho\right)^2 \tag{10.37}$$

from (10.36).
   (iii) If $i = k$ or $j = l$ but not both, we have a type which may be evaluated by considering the case $i = k = 1, j = 2, k = 3$. Writing a single integral sign for convenience, let

$$M = E(\text{sgn } u_{12} \text{ sgn } v_{12} \text{ sgn } u_{13} \text{ sgn } v_{13})$$

$$= \frac{1}{\pi^2} \int \frac{dt_1 \, dt_2 \, dt_3 \, dt_4}{it_1 \; it_2 \; it_3 \; it_4} \int fe^{i\Omega} \, dx_1 \, dx_2 \, dx_3 \, dy_1 \, dy_2 \, dy_3, \tag{10.38}$$

where we now write (dropping $\sqrt{2}$) $u_{12} = x_1 - x_2$, etc., and thus

$$\Omega = (x_1 - x_2)t_1 + (y_1 - y_2)t_2 + (x_1 - x_3)t_3 + (y_1 - y_3)t_4$$

$$= x_1(t_1 + t_3) - x_2 t_1 - x_3 t_3 + y_1(t_2 + t_4) - y_2 t_2 - y_3 t_4. \tag{10.39}$$

For the integration of $fe^{i\Omega}$ over the values of x and y we may use the known properties of characteristic functions or integrate directly to find

$$T = \int fe^{i\Omega} \, dx_1 \cdots dy_1 \cdots = \exp\{-\tfrac{1}{2}[(t_1 + t_3)^2 + (t_2 + t_4)^2$$

$$+ t_1^2 + t_2^2 + t_3^2 + t_4^2] + 2\rho'[(t_1 + t_3)(t_2 + t_4) + t_1 t_2 + t_3 t_4]\}$$

and thus

$$\frac{\partial T}{\partial \rho'} = -T(2t_1 t_2 + 2t_3 t_4 + t_1 t_4 + t_2 t_3). \tag{10.40}$$

If we differentiate M of (10.38) with respect to ρ and substitute from (10.40) we have an expression

$$\frac{\partial M}{\partial \rho} = \frac{2}{\pi^4} \int dt_1 \, dt_2 \, \frac{dt_3 \, dt_4}{it_3 \, it_4} \, T + \frac{2}{\pi^4} \int dt_3 \, dt_4 \, \frac{dt_1 \, dt_2}{it_1 \, it_2} \, T$$

$$+ \frac{1}{\pi^4} \int dt_1 \, dt_4 \, \frac{dt_2 \, dt_3}{it_2 \, it_3} \, T + \frac{1}{\pi^4} \int dt_2 \, dt_3 \, \frac{dt_1 \, dt_4}{it_1 \, it_4} \, T.$$

Because of the symmetry of T and t_1 and t_3, t_2 and t_4, this may be reduced to

$$\frac{\partial M}{\partial \rho} = \frac{4}{\pi^4} \int dt_1 \, dt_2 \, \frac{dt_2 \, dt_4}{it_3 \, it_4} \, T + \frac{2}{\pi^4} \int dt_1 \, dt_4 \, \frac{dt_2 \, dt_3}{it_2 \, it_3} \, T. \qquad (10.41)$$

We now carry out the integration in the first part with respect to t_1 and t_2, obtaining

$$\int dt_1 \, dt_2 \, T = \frac{\pi}{\sqrt{(1 - \rho^2)}} \exp[-\tfrac{3}{4}(t_3^2 + 2\rho t_3 t_4 + t_4^2)].$$

The remaining part of the integration can now be completed in the manner which led to the evaluation of (10.33).

Similarly the second integral in (10.41) may be evaluated. We obtain

$$\frac{\partial M}{\partial \rho} = \frac{8 \sin^{-1} \rho}{\pi^2 \sqrt{(1 - \rho^2)}} - \frac{4}{\pi^2} \frac{\sin^{-1} \tfrac{1}{2}\rho}{\sqrt{[1 - (\tfrac{1}{2}\rho)^2]}}. \qquad (10.42)$$

When $\rho = 1$, $M = 1$, and by a further integration we find

$$M = \left(\frac{2}{\pi} \sin^{-1} \rho\right)^2 - \left(\frac{2}{\pi} \sin^{-1} \tfrac{1}{2}\rho\right)^2 + \frac{1}{9}. \qquad (10.43)$$

There are $\binom{n}{2}$ cases of type (i), $\binom{n}{2}\binom{n-2}{2}$ cases of type (ii), and $6\binom{n}{3}$ cases of type (iii). Thus

$$E(t^2) = \frac{1}{\binom{n}{2}^2} \left\{ \binom{n}{2} + \binom{n}{2}\binom{n-2}{2}\left(\frac{2}{\pi} \sin^{-1} \rho\right)^2 \right.$$

$$\left. + 6\binom{n}{3}\left[\frac{1}{9} + \left(\frac{2}{\pi} \sin^{-1} \rho\right)^2 - \left(\frac{2}{\pi} \sin^{-1} \tfrac{1}{2}\rho\right)^2\right]\right\}. \qquad (10.44)$$

Subtracting the square of $E(t)$, we find, after a little rearrangement,

$$\operatorname{var} t = \frac{1}{\binom{n}{2}} \left\{1 - \left(\frac{2}{\pi} \sin^{-1} \rho\right)^2 + 2(n - 2)\left[\frac{1}{9} - \left(\frac{2}{\pi} \sin^{-1} \tfrac{1}{2}\rho\right)^2\right]\right\}, \qquad (10.45)$$

which is the result given in (9.13).

10.14   We may approximate to the formula in (10.45) as follows. If

$$\sin^{-1} \tfrac{1}{2}\rho = \alpha$$

$$\tfrac{1}{3}\sin^{-1}\rho = \beta,$$

then

$$2\sin\alpha = \sin 3\beta = 3\sin\beta(1 - \tfrac{4}{3}\sin^2\beta).$$

But

$$|\beta| \le \frac{\pi}{6}$$

and hence

$$1 - \tfrac{4}{3}\sin^2\beta \ge \tfrac{2}{3}.$$

Thus

$$|\alpha| \ge |\beta|$$

and hence

$$0 \le \tfrac{1}{9} - \left(\frac{2}{\pi}\sin^{-1}\tfrac{1}{2}\rho\right)^2 \le \tfrac{1}{9}\left[1 - \left(\frac{2}{\pi}\sin^{-1}\rho\right)^2\right] \le \tfrac{4}{9}pq \qquad (10.46)$$

where p, q are defined in Section 9.11. Furthermore

$$\frac{2}{n(n-1)}[1 + \tfrac{2}{9}(n-2)] \le \frac{5}{9(n-1)}, \qquad n \ge 10. \qquad (10.47)$$

Using these results in (10.45) we find

$$\operatorname{var} t \le \frac{20pq}{9(n-1)}$$

$$= \frac{5(1-t^2)}{9(n-1)}, \qquad (10.48)$$

as given in (9.18).

10.15   The remaining results we require are proved by the same method as the one used to derive $E(t)$ and var t; but the detailed working rapidly becomes more complicated.

Consider first of all $E(r_s)$. As in (2.35) we have

$$E(r_s) = \frac{3}{n^3 - n}E(\Sigma\, a_{ij}b_{ij} + \Sigma\, a_{ij}b_{ik}), \qquad j \ne k$$

$$= \frac{3}{n^3 - n}[n(n-1)E(a_{ij}b_{ij}) + n(n-1)(n-2)E(a_{ij}b_{ik})]$$

$$= \frac{3}{n+1}E(a_{ij}b_{ij}) + \frac{3(n-2)}{n+1}E(a_{ij}b_{ik}). \qquad (10.49)$$

We have already found that

$$E(a_{ij}b_{ij}) = \frac{2}{\pi}\sin^{-1}\rho.$$

To evaluate $E(a_{ij}b_{ik})$ consider the case $i = 1$, $j = 2$, $k = 3$. In the manner of Section 10.11 we find

$$E(a_{12}b_{13}) = \frac{1}{\pi^2}\int_{-\infty}^{\infty}\int_{-\infty}^{\infty}\frac{dt_1}{it_1}\frac{dt_2}{it_2}E\{\exp[it(x_1 - x_2) + it_2(y_1 - y_2)]\}$$

$$= \frac{1}{\pi^2}\int_{-\infty}^{\infty}\int_{-\infty}^{\infty}\frac{dt_1}{it_1}\frac{dt_2}{it_2}\exp[-\tfrac{1}{2}(t_1^2 + \rho t_1 t_2 + t_2^2)]. \qquad (10.50)$$

This is like the integral of Section 10.12 with ρ instead of 2ρ and is therefore $2/\pi \sin^{-1}\tfrac{1}{2}\rho$. Hence we find

$$E(r_s) = \frac{6}{\pi(n + 1)}[\sin^{-1}\rho + (n - 2)\sin^{-1}\tfrac{1}{2}\rho]. \qquad (10.51)$$

10.16   The detailed evaluation of var r_s and cov(r_s, t) proceeds in the same manner, but in the former we arrive at integrals of the elliptic type. Recourse must then be had to expansions in powers of ρ, and this leads to equations (9.30) and (9.31). Reference may be made to Kendall (1949) and David, Kendall and Stuart (1951) for the details.

Notes and references

See references to the previous chapter. Kendall (1949) considered the relationship between rank coefficients and parent parameters in non-normal populations expressed as a Gram–Charlier series. Sundrum (1953b) obtained the third and fourth moments of t and considered their evaluation in the normal case.

Fieller, Hartley and Pearson (1957) obtained some improved formulae for the asymptotic variance of r_s. David and Mallows (1961) gave some exact expressions which were tabulated by Fieller and Pearson (1961).

Chapter 11

Paired comparisons

11.1 Up to this stage we have considered rankings as given by the circumstances of the problem, and have not concerned ourselves with the question of whether the data properly lend themselves to a ranking treatment. Cases often arise, particularly in psychological work, where there is some doubt on this point. Suppose we ask an observer to rank n men in order of intelligence. He may attempt to do so, and may even succeed in producing a ranking, although the nature of 'intelligence' is so obscure that we cannot assume the possibility of ordering individuals by reference to it. To some extent we are begging the question by assuming that intelligence is a linear variable. Again, we may ask an observer to rank a number of districts according to preference for living in them; but the preferences will depend on a number of factors such as cost, availability of transport, height above sea level, or nearness to shopping centres, and it by no means follows that the observer is capable of expressing a final preference on a linear scale. If we insist on a ranking we may be forcing the data, so to speak, into an over-narrow framework which will distort the true situation. The method we shall discuss in this section is designed to overcome such difficulties.

11.2 We shall suppose that of n objects each of the possible $\frac{1}{2}n(n-1)$ pairs is presented to an observer, one pair at a time, and that preference is recorded for one member of the pair. If A is preferred to B we may write $A \rightarrow B$ or $B \leftarrow A$.

Example 11.1

In some experiments on a dog, six different kinds of food were prepared. Each of the $\binom{6}{2} = 15$ possible pairs of kinds of food was offered to the dog and a note was made of which kind was taken first. If we denote the six foods by the letters A to F we may record the results as in Table 11.1.

For instance an entry 1 in column B and row A means $A \rightarrow B$ and, of course, corresponds to 0 in row B and column A. Thus, in the above table, $A \rightarrow B$, $A \rightarrow C$, $A \leftarrow D$, etc. The diagonals are blocked out.

The arrangement of the objects in rows and columns is arbitrary, but it is clearly convenient to have the orders in row and column the same.

Table 11.1 Preferences of a dog for six foods

	A	B	C	D	E	F
A	–	1	1	0	1	1
B	0	–	0	1	1	0
C	0	1	–	1	1	1
D	1	0	0	–	0	0
E	0	0	0	1	–	1
F	0	1	0	1	0	–

The complex of preferences may also be represented diagrammatically. We represent the six objects *A* to *F* by the vertices of a regular polygon as in Fig. 11.1. The vertices are joined in all possible ways by straight lines and if *A* → *B* we draw an arrow on the line *AB* pointing from *A* to *B*.

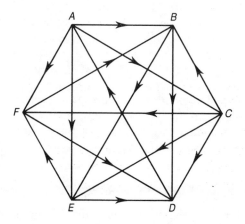

Fig. 11.1.

11.3 If an observer expresses preferences for three objects *A*, *B*, *C* as *A* → *B* → *C* → *A* or *A* ← *B* ← *C* ← *A* we shall say that the triad is 'circular' or 'inconsistent'. In the triangle *ABC* all the arrows in the diagram of the type in Table 11.1 go round the same way. In Fig. 11.1 the triads *ACD*, *BEF* and three others are circular.

Clearly circular triads cannot arise in ordinary ranking for if *A* → *B* and *B* → *C* then *A* → *C*. It is then a necessary and sufficient condition for the possibility of expressing the preferences as a ranking that no circular triads be present. The more circular triads there are, so to speak, the further we depart from the ranking situation towards a position of inconsistency under which *A* may be preferred to *B* and *B* to *C* but nevertheless *C* is preferred to *A*.

11.4 It is possible to have circular polyads of extent greater than three. For instance if *A* → *B* → *C* → *D* → *A* the tetrad *ABCD* is circular. A circular *n*-ad must, however, contain *n* − 2 circular triads but it may contain more; and the fact that it contains circular triads does not imply that it is itself circular.

Suppose, for instance, that $ABCD$ is circular. Then either $A \rightarrow C$ or $C \rightarrow A$. In the first case ACD is circular, in the second ABC. Similarly either ABD or BCD is circular. Thus the tetrad must contain at least two circular triads. On the other hand, the scheme expressed by $A \rightarrow B \rightarrow C \leftarrow D \rightarrow A$, $B \rightarrow D$, $C \rightarrow A$ contains the circular triad ABC and ABD, but $ABCD$ is not circular.

We shall therefore concentrate on circular triads which compose the elementary inconsistencies of the situation, and shall ignore the more ambiguous criteria based on polyads of greater extent.

11.5   It will be shown in the next chapter that if n is odd the maximum number of circular triads is $\frac{1}{24}(n^3 - n)$, and if n is even the maximum number is $\frac{1}{24}(n^3 - 4n)$. The minimum number is zero. We may therefore define a *coefficient of consistence* by the equations

$$\zeta = 1 - \frac{24d}{n^3 - n}, \qquad n \text{ odd}$$

$$= 1 - \frac{24d}{n^3 - 4n}, \qquad n \text{ even},$$

(11.1)

where d is the observed number of circular triads. If and only if ζ is unity there are no circular triads and the data may be ranked.

For example, in the data of Example 11.1 there are 5 circular triads. The maximum number is 8, so $\zeta = 0.375$.

11.6   We may, in a certain sense, test the significance of a value of ζ by considering the distribution it would have if all the preferences were allotted at random. This will tell us whether the observed ζ could have arisen by chance if the observer was completely incompetent, or, alternatively, whether there is some degree of consistency in his preferences notwithstanding a lack of perfection.

Appendix Table 9 gives the probablities that certain values of d will be attained or exceeded for $n = 2$ to 7, on the assumption that preferences are allotted purely at random so that any preference scheme is as probable as any other. These distributions are rather troublesome to obtain, and in practice are not often required for larger values of n; but when a test is required it may be derived from the χ^2-distribution, to which that of d tends as n increases. In fact, writing

$$v = \frac{n(n - 1)(n - 2)}{(n - 4)^2}$$

$$\chi^2 = \frac{8}{n - 4}\left[\frac{1}{4}\binom{n}{3} - d + \tfrac{1}{2}\right] + v.$$

(11.2)

We have the result that χ^2 is distributed approximately in the usual form with v degrees of freedom. The distribution is, however, measured from higher to lower values of d, so that the probability that d will be attained or exceeded is the complement of the probability for χ^2. The following example illustrates the point.

Example 11.2

In a set of 7 a value of d equal to 13 is observed. From (11.2) we have

$$v = \frac{7(6)(5)}{9} = 23.33$$

$$\chi^2 = \tfrac{8}{3}(8.75 - 13 + \tfrac{1}{2}) + 23.33 = 13.33.$$

From Appendix Table 8 we see that these values correspond approximately to a significance level of about 0.95. The appropriate significance level for d is thus $1 - 0.95 = 0.05$. The exact value of the probablity, from Appendix Table 9, is 0.036. The approximation is fair for such a low value of n.

11.7   In a table of the type of 11.1 it is possible to ascertain the number of circular triads d without counting them directly. Suppose the row totals are $a_1, \ldots, a_n$. Then

$$d = \tfrac{1}{6}n(n - 1)(n - 2) - \tfrac{1}{2} \sum_{i=1}^{n} a_i(a_i - 1). \qquad (11.3)$$

For example, in Table 11.1 the row totals are 4, 2, 4, 1, 2, 2, totalling $15 = \binom{6}{2}$. Thus

$$\tfrac{1}{2} \sum a(a - 1) = \tfrac{1}{2} \sum a^2 - \tfrac{1}{2} \sum a$$

$$= \tfrac{1}{2}(45 - 15) = 15$$

giving

$$d = 20 - 15 = 5.$$

The same formula applies to column totals, say $b_1, \ldots, b_n$. For we have, in view of the method of construction of the tables,

$$b_i = (n - 1) - a_i.$$

Thus

$$\sum b_i = \sum a_i$$

$$\sum b_i^2 = n(n - 1)^2 - 2(n - 1) \sum a + \sum a^2$$

$$= \sum a_i^2$$

and hence

$$\sum b(b - 1) = \sum a(a - 1).$$

We can also write (11.3) in the form

$$d = \tfrac{1}{12}n(n - 1)(2n - 1) - \tfrac{1}{2} \sum a_i^2, \qquad (11.4)$$

which is probably the simplest for calculation.

Coefficient of agreement

11.8   Suppose now that we have m observers each of which provides $\binom{n}{2}$ preferences between pairs of n objects. Suppose that in a table of the form in Table 11.1 we enter a unit in the cell in row A and column B whenever $A \rightarrow B$ and count the units in each cell. A cell may then contain any number from 0 to m. If the observers are in complete agreement there will be $\binom{n}{2}$ cells containing m, the remaining cells being zero. The agreement may be complete even if there are inconsistencies present.

Suppose that the cell in row A and column B contains the number y. Let

$$\Sigma = \Sigma \binom{y}{2}, \qquad (11.5)$$

the summation extending over the $n(n-1)$ cells of the table (the diagonal cells being ignored). Then Σ is the sum of the number of agreements between pairs of judges. Define

$$u = \frac{2\Sigma}{\binom{m}{2}\binom{n}{2}} - 1$$

$$= \frac{8\Sigma}{m(m-1)n(n-1)} - 1. \qquad (11.6)$$

We shall call u the *coefficient of agreement*. If there is complete agreement, and only in this case, $u = 1$. The further we depart from this case, as measured by agreements between pairs of observers, the smaller u becomes. The minimum number of agreements in each cell is $\frac{1}{2}m$ if m is even or $\frac{1}{2}(m \pm 1)$ if m is odd. In the first case the value of u is $-1/(m-1)$, in the second $-1/m$.

This minimum value is not -1 unless $m = 2$. In such a case, with two observers, we have

$$u = \frac{2\Sigma}{\frac{1}{2}n(n-1)} - 1, \qquad (11.7)$$

so that u may be regarded as a generalisation of the coefficient tau in this case.

Example 11.3

Students in a class of boys (ages 11 to 13 inclusive) were asked to state their preferences with respect to certain school subjects. Each child was given a sheet on which were written the possible pairs of subjects and asked to underline the one preferred in each case. The preferences of 21 boys in 13 school subjects are shown in Table 11.2.

The calculation of Σ for this table, in which the objects are arranged in order of total number of preferences, may be shortened by noting that Σ, as

Table 11.2 Preferences of 21 boys in 13 subjects

	1	2	3	4	5	6	7	8	9	10	11	12	13	Totals
1. Woodworking	—	14	20	15	15	16	16	18	18	18	20	21	20	211
2. Gymnastics	7	—	14	12	13	18	14	16	16	20	16	18	19	183
3. Art	1	7	—	10	14	10	16	18	16	16	17	16	19	160
4. Science	6	9	11	—	11	12	15	14	13	13	17	17	16	154
5. History	6	8	7	10	—	14	11	12	14	15	13	14	16	140
6. Geography	5	3	11	9	7	—	14	14	13	13	16	15	17	137
7. Arithmetic	5	7	5	6	10	7	—	9	11	13	15	13	15	116
8. Religion	3	5	3	7	9	7	12	—	12	14	14	16	14	116
9. English literature	3	5	5	8	7	8	10	9	—	10	13	13	15	106
10. Commercial subjects	3	1	5	8	6	8	8	7	11	—	10	10	14	91
11. Algebra	1	5	4	4	8	5	6	7	8	11	—	10	13	82
12. English grammar	0	3	5	4	7	6	8	5	8	11	11	—	13	81
13. Geometry	1	2	2	5	5	4	6	7	6	7	8	8	—	61
													Total	1638

given by equation (11.5), may be transformed into

$$\Sigma = \Sigma(\gamma)^2 - m\,\Sigma(\gamma) + \binom{m}{2}\binom{n}{2},$$

where the summation now takes place over the half of the table below the diagonal. Since the numbers in this half are smaller than those in the other half there is a considerable saving in arithmetic.

We find

$$\Sigma = 9718$$

and hence

$$u = \frac{2(9718)}{\binom{21}{2}\binom{13}{2}} - 1 = 0.186.$$

There is thus a certain amount of agreement among the children, indicated by the positive value of u.

The distribution of circular triads is shown in Table 11.3.

Table 11.3 Distribution of circular triads

No. of triads	Frequency	No. of triads	Frequency
0	1	12	1
1	1	17	3
4	5	21	1
6	2	25	1
7	2	29	1
8	1	39	1
10	1		
		Total	21

The total number of circular triads was 242 with a mean of 11.5. Only one boy was entirely consistent. On the other hand, for $n = 13$ the maximum number of circular triads is 91, with a mean value of 71.5. It is thus clear that, except perhaps for one boy, we cannot suppose that any boy allotted preferences at random. We are again led to conclude that the boys are genuinely capable of making distinctions, and that consistently, on the whole, half the boys have coefficients ζ greater than 0.92.

Example 11.4

It frequently happens in practice that observers decline to express a preference between some pairs of objects. We then arrive at difficulties similar to those arising from tied ranks. We shall deal with them in Table 11.1 by putting $\frac{1}{2}$ in each of the cells row A, column B and row B, column A where no preference is expressed between A and B. The following data from J. W. Whitfield will illustrate the method.

Forty-six workers in a department were asked to say which of a pair they considered more important in the 66 possible pairs from the following items:

Ventilation	Good opportunities for promotion
Canteen facilities	Lavatories and cloakrooms
Responsibility	Work which requires no thought
Pension fund	Lighting
Interesting work	Hours of work
Security of employment (i.e. of work in general)	
Tenure of employment (i.e. at this particular factory)	

Table 11.4 shows the results.

From the sums of squares of items in Table 11.4, we find

$$\Sigma \, \gamma^2 = 86\,392, \qquad \Sigma \, \gamma = 3036.$$

Table 11.4 Data on Workers

	V	C	R	P	Op	L&C	WNT	Li	IW	HW	SE	TE	Totals
V	–	14	10	10	20	16	3	20	20	24	28½	27	192½
C	32	–	24½	26½	30½	25	0	35	28	30	34	32½	298
R	36	21½	–	21	40	33	0	35½	36	32	37	28	320
P	36	19½	25	–	31½	26	2	32	32	29	32	29½	294½
Op	26	15½	6	14½	–	23	1	27	28½	26	25½	23	216
L&C	30	21	13	20	23	–	0	18	22½	22	24	30½	224
WNT	43	46	46	44	45	46	–	46	46	44	46	44½	496½
Li	26	11	10½	14	19	28	0	–	26	25	27½	20	207
IW	26	18	10	14	17½	23½	0	20	–	14	33	23	199
HW	22	16	14	17	20	24	2	21	32	–	32	18½	218½
SE	17½	12	9	14	20½	22	0	18½	13	14	–	26½	167
TE	19	13½	18	16½	23	15½	1½	26	23	27½	19½	–	203
Totals	313½	208	186	211½	290	282	9½	299	307	287½	339	303	3036

Thus

$$\tfrac{1}{2} \sum \gamma(\gamma - 1) = 41\,678.$$

Hence, if we use formula (11.6) without regard to fractional γs,

$$u = \frac{2(41\,678)}{\binom{46}{2}\binom{12}{2}} - 1$$

$$= 0.220.$$

Consider now a score such as that in row C column R, $24\tfrac{1}{2}$, with the complementary score in column C row R of $21\tfrac{1}{2}$. As we have just calculated the score, the contributions of these two are

$$\binom{24\tfrac{1}{2}}{2} + \binom{21\tfrac{1}{2}}{2} = \frac{1}{2}\left[\binom{25}{2} + \binom{24}{2}\right] + \frac{1}{2}\left[\binom{22}{2} + \binom{21}{2}\right] - \frac{1}{4}.$$

Thus our crude method is equivalent to taking an average of the undetermined preference, first by assuming $R \rightarrow C$ so that the scores are 25 and 21, secondly by assuming $R \leftarrow C$ so that the scores are 24 and 22; except for the factor $\tfrac{1}{4}$ which is negligible. This is in accordance with our treatment of ties in the ranking case.

In Table 11.4 the score shown for row R column P, 21, was in fact $20\tfrac{2}{3}$, i.e. comprised two halves, and similarly that in column R row P was $24\tfrac{2}{3}$. If we were to average the possible scores we should have

$$\tfrac{1}{4}(20, 26) + \tfrac{1}{2}(21, 25) + \tfrac{1}{4}(22, 24)$$

and the difference from our actual count of (21, 25) is thus

$$\frac{1}{8}\left[\binom{20}{2} + \binom{26}{2} + \binom{22}{2} + \binom{24}{2} - 2\binom{21}{2} - 2\binom{25}{2}\right] = \frac{1}{2}.$$

Again the difference is negligible.

11.9   A test of significance of u can be obtained by considering what the distribution would be if all the preferences were allotted at random. These distributions have been worked out for values of $m = 3$, $n = 2$ to 8; $m = 4$, $n = 2$ to 6; $m = 5$, $n = 2$ to 5; $m = 6$, $n = 2$ to 4, and form the basis of Appendix Tables 10.

For higher values of m and n an adequate approximation is given by the χ^2-distribution. We write

$$\chi^2 = \frac{4}{m-2}\left[\Sigma - \frac{1}{2}\binom{n}{2}\binom{m}{2}\frac{m-3}{m-2}\right] \tag{11.8}$$

$$v = \binom{n}{2}\frac{m(m-1)}{(m-2)^2} \tag{11.9}$$

and test significance using the χ^2-distribution with v degrees of freedom. A continuity correction may be applied by deducting unity from Σ.

For example, with $m = 3$, $n = 8$, we have

$$\chi^2 = 4\,\Sigma, \qquad v = 168.$$

From Appendix Table 10A we have, exactly,

$$\text{for } \Sigma = 54, \qquad P = 0.011$$

$$\text{for } \Sigma = 58, \qquad P = 0.0011.$$

For these values of Σ (with continuity corrections) the corresponding values of χ^2 are 212 and 228. For $v = 168$ we can take $\sqrt{(2\chi^2)}$ to be normally distributed with $\sqrt{(2v - 1)} = 18.30$ and variance 1, so that our deviates are

$$\sqrt{[(2)(212)]} - 18.30 = 2.29$$

and

$$\sqrt{[(2)(228)]} - 18.30 = 3.05.$$

These are seen from Appendix Table 3 to correspond to probabilities 0.011 and 0.00114, very close to the exact values.

Similarly, with $m = 6$, $n = 4$ we find from (11.8) and (11.9) that $\Sigma - 33.75$ is distributed with 11.25 degrees of freedom. From Appendix Table 10D we see that the 1 per cent point lies between $\Sigma = 59$ and $\Sigma = 60$. The corresponding χ^2 values are then (with continuity corrections) 24.25 and 25.25. From Appendix Table 8 we see that these values do in fact fall very close to the 1 per cent point for $v = 11.25$, which is somewhere between 24.725 ($v = 11$) and 26.217 ($v = 12$).

Example 11.5

In Example 11.3, for $n = 13$, $m = 21$ we found $\Sigma = 9718$ and $u = 0.186$. This indicates some community of preference but not a very large amount. Is the value significant?

From (11.8) and (11.9) we find

$$\chi^2 = \frac{4}{21}\left\{ 9718 - \frac{1}{2} \cdot \frac{18}{19} \binom{13}{2}\binom{21}{2} \right\} = 421.4$$

$$v = \binom{13}{2}\frac{21(20)}{19^2} = 90.7$$

$$\sqrt{(2\chi^2)} - \sqrt{(2v - 1)} = 15.3.$$

This is far beyond any ordinary significance point, and we conclude that the observed u could not have arisen by chance from a population in which all the boys allotted preferences at random. This confirms our conclusion reached in Example 11.3.

Again, in Example 11.4 we find

$$\chi^2 = 754.5, \qquad v = 70.6$$

$$\sqrt{(2\chi^2)} - \sqrt{(2v - 1)} = 27.02,$$

again an extremely improbable result if the preference were allotted at random. We may conclude that the observed value of u is significant.

11.10 Ehrenberg (1952) corrects a result of Kendall and Smith (1940) concerning a test of significance of u when the data are ranked. If $M = \binom{m}{2}$ and $N = \binom{n}{2}$, u may be tested using the chi-square distribution with

$$\chi^2 = \frac{6(2n + 5)MN}{(m - 2)(2n^2 - 6n + 17)} u + v$$

where

$$v = \frac{2(2n + 5)^2 MN}{(m - 2)^2(2n^2 + 6n + 7)^2}.$$

11.11 In the paired-comparison case, models of considerable generality can be written down, although their practical application is not easy to make. For instance, suppose that we have n objects and the probability of ranking the ith higher than the jth is π_{ij}. This model would be appropriate to a situation where an individual made preferences which were subject to error, or to the case where a number of individuals differed in their preferences. The coefficient of agreement u defined in (11.6) then has mean

$$E(u) = 1 - \frac{4}{m(m - 1)} \sum_{i<j}^{n} \pi_{ij}(1 - \pi_{ij}) \tag{11.10}$$

and variance

$$\text{var } u = \frac{64}{m(m - 1)n^2(n - 1)^2} \sum_{i<j} \{(m - 1)[\pi_{ij}(1 - \pi_{ij})] - 4\pi_{ij}^2(1 - \pi_{ij})^2\}. \tag{11.11}$$

The difficulty in applying the result resides in our ignorance of the πs.

11.12 An approach of a different kind is to suppose that n objects have 'true' ratings, expressed by the numbers $\pi_1, \ldots, \pi_n$ ($\sum_{j=1}^{n} \pi_j = 1$) such that, when the ith and jth objects are compared, the probability that the ith is preferred to the jth is $\pi_i/(\pi_i + \pi_j)$. This model, which has a certain prior plausibility for much paired comparison work, is simpler than the previous one, since it depends on n parameters π_i instead of $\frac{1}{2}n(n - 1)$ parameters π_{ij}. Bradley and Terry have developed methods of estimating the πs from observation and of testing the significance of a set of results.

Notes and references

The original references on paired comparisons are Kendall and Smith (1940) and Moran (1947), and for the non-null case, Smith (1950), Ehrenberg (1952), and Mallows (1957). The model developed by Mallows was extended by Feigin and Cohen (1978) to provide an estimate of the underlying ranking and procedures for inference. Other early references are Guttman (1946), Wei (1952), Kendall (1955), Bose (1956), Wilkinson (1957), Slater (1961), and Starks and David (1961).

The model discussed in Section 11.11 was proposed by Smith (1950) and discussed by Ehrenberg (1952). The model of Section 11.12 was developed by Bradley and Terry (1952), and Bradley (1954, 1955). These papers give tables required for practical applications. Bradley (1976) categorised and discussed various model formulations, and Bradley (1984) gave a good overview with extensive references.

The literature on this general topic is vast. A bibliography published by Davidson and Farquahr (1976) has about 400 references. The most recent and complete treatment is provided in the revised and expanded monograph by David (1988).

Chapter 12

Proof of the results of Chapter 11

12.1 We will first establish the result that in a complex of paired comparisons the maximum number of circular triads is $\frac{1}{24}(n^3 - n)$ for n odd and $\frac{1}{24}(n^3 - 4n)$ for n even.

Consider a polygon of the type of Fig. 11.1 with n vertices. There will be $n - 1$ lines emanating from each vertex. Let $a_1, \ldots, a_n$ be the number of arrows which *leave* the vertices. Then

$$\sum_{j=1}^{n} a_j = \binom{n}{2} \tag{12.1}$$

and the mean value of a is $\frac{1}{2}(n - 1)$. Consider the function

$$T = \sum_{j=1}^{n} [a_j - \tfrac{1}{2}(n - 1)]^2, \tag{12.2}$$

that is to say, n times the variance of the a-numbers. We have at once

$$T = \sum a_j^2 - \tfrac{1}{4}n(n - 1)^2. \tag{12.3}$$

12.2 We now show that if the direction of a preference is altered and the effect is to increase the number of circular triads by p, T is reduced by $2p$ and vice versa. Consider the preference $A \rightarrow B$. The only triads affected by reversing to $B \rightarrow A$ are those containing the line AB. Suppose there are α preferences of type $A \rightarrow X$ (including $A \rightarrow B$) and β of type $B \rightarrow X$. Then there are four possible types of triads:

$$A \rightarrow X \leftarrow B, \quad \text{say } x \text{ in number}$$

$$A \leftarrow X \rightarrow B,$$

$$A \rightarrow X \rightarrow B, \quad \text{which must number } \alpha - x - 1$$

$$A \leftarrow X \leftarrow B, \quad \text{which must number } \beta - x.$$

When the preference $A \rightarrow B$ is reversed the first two remain non-circular. The third becomes circular, the fourth ceases to be so. The increase in the number of circular triads is

$$(\alpha - x - 1) - (\beta - x) = \alpha - \beta - 1 = p.$$

The reduction in T is

$$\alpha^2 - (\alpha - 1)^2 + \beta^2 - (\beta + 1)^2 = 2(\alpha - \beta - 1) = 2p.$$

Our result follows; for the effect of altering individual preferences is cumulative on T and d.

12.3   From the definition of T it is clear that the maximum value occurs when the data are ranked, and thus $\max T = \frac{1}{12}(n^3 - n)$. For the minimum value consider a polygon with vertices $A_1, \ldots, A_n$. Set up the preferences $A_1 \to A_2 \to \cdots \to A_n$. Next set up the preferences $A_1 \to A_3 \to A_5 \to \cdots$. If this does not provide a closed tour of all the points of the polygon proceed to the next unvisited vertex, A_k, and set up the preferences $A_k \to A_{k+2} \to$ etc., and so on. Then set up the preferences $A_1 \to A_4 \to A_7$, etc., and so on until the whole preference schema is completed.

If n is odd the preferences described will consist of circular tours of the polygon, and each a will be $\frac{1}{2}(n - 1)$ so that $T = 0$; and this is obviously a minimum. If n is even the last preference $A_1 \to A_{(1/2)n+1}$ will not be a tour but will consist of a single line joining one vertex with the symmetrically opposite vertex. Thus there will be $\frac{1}{2}n$ vertices with $a = \frac{1}{2}n$ and $\frac{1}{2}n$ with $a = \frac{1}{2}n - 1$. In this case $T = \frac{1}{4}n$ and again this is a minimum.

Thus T can range from 0 or $\frac{1}{4}n$ to $\frac{1}{12}(n^3 - n)$ and since an increase of two in T corresponds to a decrease of unity in d, there are the following maximum values of d:

$$\frac{1}{24}(n^3 - n), \qquad n \text{ odd},$$

$$\frac{1}{24}(n^3 - 4n), \qquad n \text{ even}.$$

12.4   The numbers a are the totals of rows in Table 11.1, and from what has been said it follows that the number of circular triads is given by

$$d = \frac{1}{2}[\frac{1}{12}(n^3 - n) - T]$$

$$= \frac{1}{2}[\frac{1}{12}(n^3 - n) + \frac{1}{4}n(n - 1)^2 - \sum a_j^2]$$

$$= \frac{1}{12}n(n - 1)(2n - 1) - \frac{1}{2}\sum a_j^2$$

$$= \frac{1}{6}n(n - 1)(n - 2) - \frac{1}{2}\sum a_j(a_j - 1),$$

as given in (11.3) and (11.4).

12.5   We now establish the χ^2-approximation to the distribution of d as n becomes large.

Let the objects be numbered from 1 to n. Write $P_{ijk} = 1$ if the triad (i, j, k) is circular and $P_{ijk} = 0$ if it is not. Then

$$d = \sum P_{ijk}, \tag{12.4}$$

where the summation is over all triads. Thus

$$E(d) = \binom{n}{3} E(P_{ijk}).$$

By enumerating the possible cases for preferences in a given triad we see that $E(P_{ijk}) = \frac{1}{4}$. Hence, for the mean value of d,

$$E(d) = \frac{1}{4}\binom{n}{3}. \tag{12.5}$$

Now consider $E(\sum P_{ijk})^2$. When we expand this there will be $\binom{n}{3}$ terms of type P_{ijk}^2; $3\binom{n}{3}\binom{n-3}{2}$ terms of type $P_{ijk}P_{ilm}$ where $j \neq l$, $k \neq m$; $3\binom{n}{3}(n-3)$ terms of type $P_{ijk}P_{ijl}$ where $k \neq l$; and $\binom{n}{3}\binom{n-3}{3}$ terms of type $P_{ijk}P_{lmn}$ with different suffixes. By examining particular configurations we see that the expectation of the first is $\frac{1}{4}$ and of each of the others is $\frac{1}{16}$. Thus

$$E(\sum P_{ijk})^2 = \frac{1}{16}\binom{n}{3}\left[4 + \frac{3}{2}(n-3)(n-4) + 3(n-3) + \binom{n-3}{3}\right]$$

$$= \frac{1}{16}\binom{n}{3}\left[\binom{n}{3} + 3\right].$$

Hence

$$\mu_2(d) = E(d - \bar{d})^2 = \frac{3}{16}\binom{n}{3}. \tag{12.6}$$

The calculation of the third and fourth moments is much more complicated, but is the same in principle. We find

$$\mu_3 = -\frac{3}{32}\binom{n}{3}(n-4) \tag{12.7}$$

$$\mu_4 = \frac{1}{55296}\binom{n}{3}(972n^3 + 972n^2 - 36936n + 80352). \tag{12.8}$$

12.6 The moments of the χ^2-distribution

$$dF \propto e^{-(1/2)x^2}\chi^{\nu-1}\,d\chi,$$

are given by

$$\mu_1' \text{ (about zero)} = \nu$$

$$\mu_2 \qquad\qquad = 2\nu$$

$$\mu_3 \qquad\qquad = 8\nu$$

We note that $\mu_3(d)$ is negative while $\mu_3(\chi^2)$ is positive. We shall therefore measure d from the other end of its distribution so as to bring the distributions into accord. Applying a correction for continuity, we see that

$$x = k\left[\frac{1}{4}\binom{n}{3} - d + \tfrac{1}{2}\right] + v \qquad (12.9)$$

has mean value v, the mean of χ^2. We will choose k so that the distributions have the same variance, which is obviously so when

$$k = \sqrt{\frac{\mu_2(\chi^2)}{\mu_2(d)}} = \sqrt{\frac{2v}{\frac{3}{16}\binom{n}{3}}}. \qquad (12.10)$$

For the third moments to be equal we must have

$$\frac{\frac{3}{32}\binom{n}{3}(n-4)}{\left[\frac{3}{16}\binom{n}{3}\right]^{3/2}} = \frac{8v}{(2v)^{3/2}}$$

leading to

$$v = \frac{n(n-1)(n-2)}{(n-4)^2}. \qquad (12.11)$$

Using this value of v in (12.10) we find that

$$\frac{8}{n-4}\left[\frac{1}{4}\binom{n}{3} - d + \tfrac{1}{2}\right] + v \qquad (12.12)$$

has the first three moments the same as those of χ^2.

12.7   It may be shown that the distribution of d tends to normality as n tends to infinity. The proof follows the lines of those given in Chapter 5. We show that moments of odd order are of lower order in n that those of even order. In an expansion

$$(\Sigma \, Q_{ijk})^{2m},$$

where $Q_{ijk} = P_{ijk} - \tfrac{1}{4}$, the dominant term is of type $Q_{ijk}^2 Q_{lmn}^2 \cdots$ occurring in m factors with expectation $\frac{3}{32}$ and frequency $(2m)!/(2^m m!)$. Thus the dominant term gives

$$\mu_{2m} \sim \frac{(2m)!}{2^m m!}\left(\frac{3}{32}\right)^m n^{3m}$$

$$\sim \frac{(2m)!}{2^m m!}(\mu_2)^m$$

and the tendency to normality follows. For details see Moran (1947).

12.8   Finally, we derive the χ^2-approximation for testing the coefficient of agreement, u, when all preferences are allotted at random.

The contribution to Σ from two cells (row A, column B) and (row B, column A) is typified by

$$\binom{\gamma}{2} + \binom{m-\gamma}{2}. \tag{12.13}$$

Of the 2^m total ways in which the m preferences can be allotted to the cells there will be $\binom{m}{\gamma}$ in which γ units occur. Consequently the frequency of the contribution to Σ is the corresponding coefficient of t in

$$f = t^{\binom{m}{2}} + \binom{m}{1}t^{\binom{m-1}{2}} + \cdots + \binom{m}{\gamma}t^{\binom{m-\gamma}{2}+\binom{\gamma}{2}} + \cdots + t^{\binom{m}{2}}. \tag{12.14}$$

Now if the preferences are allotted at random the contributions to Σ for the $\binom{n}{2}$ cells are independent; and hence the distribution of Σ is given by

$$f^{\binom{n}{2}}, \tag{12.15}$$

the frequency of Σ being the coefficient of t^Σ in (12.15).
For instance, with $m = 3$, $n = 4$.

$$f = t^3 + 3t + 3t + t^3 = 2t(3 + t^2)$$

and the distribution is arrayed by

$$\{2t(3 + t^2)\}^6 = 2^6 t^6(729 + 1458t^2 + 1215t^4 + 540t^6 + 135t^8 + 18t^{10} + t^{12})$$

with a total frequency of $2^6(4^6)$. These and similar values form the basis of Appendix Tables 10.

12.9   For constant m the distribution of χ^2 tends to normality with increasing n, for it is the mean of $\binom{n}{2}$ constituents with finite equal moments.* For constant n the distribution tends with increasing m to a form of the χ^2-distribution; for each of the $\binom{n}{2}$ cells contributes a variate $\binom{\gamma}{2} + \binom{m-\gamma}{2}$ which is effectively like γ^2, and the distribution of γ tends to normality.
The rth moment of Σ about the origin is given by

$$2^{m\binom{n}{2}}\mu_r' = \left[\left(t\frac{\partial}{\partial t}\right)^r f^{\binom{n}{2}}\right]_{t=1}. \tag{12.16}$$

The differentiation, in fact, multiplies a term containing t^Σ by Σ^r, and when we put $t = 1$ the array becomes the sum of frequencies each multiplied by Σ^r

---

* This is a particular case of the central limit theorem—see Kendall and Stuart, *Advanced Theory*, Vol. 1, 7.26.

which provides the rth moment. Thus, for the first moment,

$$2^m \mu_1' = \binom{n}{2} \sum_{j=0}^{m} \binom{m}{j} [j^2 - mj + \tfrac{1}{2}(m^2 - m)]$$

$$= \binom{n}{2} \left[2^m \binom{m}{2} + \Sigma \binom{m}{j} (j^2 - jm) \right]$$

giving

$$\mu_1' = \frac{1}{2} \binom{m}{2} \binom{n}{2}. \tag{12.17}$$

In a similar way—we omit the algebra—we find

$$\mu_2 = \frac{1}{4} \binom{n}{2} \binom{m}{2} \tag{12.18}$$

$$\mu_3 = \frac{3}{4} \binom{n}{2} \binom{m}{3} \tag{12.19}$$

$$\mu_4 = \binom{m}{2} \binom{n}{2} \left[\frac{3m^2 - 15m + 17}{8} + \frac{3}{32} \binom{n}{2} m(m-1) \right]. \tag{12.20}$$

Proceeding in the manner of Section 12.6 we see that

$$\chi^2 = \frac{4}{m-2} \left[\Sigma - \frac{1}{2} \binom{n}{2} \binom{m}{2} \frac{m-3}{m-2} \right] \tag{12.21}$$

is distributed in the usual form with

$$v = \binom{n}{2} \frac{m(m-1)}{(m-2)^2} \tag{12.22}$$

degrees of freedom.

Notes and references

See notes and references to the previous chapter.

Chapter 13

Some further applications

This chapter provides a guide for additional reading on topics which are related to some of the methods presented elsewhere in this book.

Estimation of population consensus

13.1 Stuart (1951) discussed the problem of sampling from a population of rankers. Suppose each member of a population arranges a number n of objects in order of preference; for example, a set of occupations may be ranked according to their social prestige. If m members are drawn at random from a population of M, we have m sets of rankings of n objects, for which we could test independence as discussed in Chapter 6.

Now, however, suppose that we want to see whether whatever community of preference as exists among the m observed members can be taken to be representative of preference in the population of M. Stuart (1951) used the concordance coefficient to discuss this problem and Linhart (1960) gave an approximate test based on average internal rank correlation. See also Ehrenberg (1952) and Hays (1960).

Two-group concordance

13.2 Tests that compare the concordance coefficient of one set of rankings with the concordance coefficient of another set of rankings are called tests for two-group concordance. Tests applicable in this situation have been developed by Schucany and Frawley (1973), Li and Schucany (1975), Schucany and Beckett (1976), Hollander and Sethuraman (1978), Schucany (1978), Beckett and Schucany (1979), Kraemer (1981), Palachek and Kerin (1982), Katz and McSweeney (1983), Costello and Wolfe (1985). An approximate test was developed earlier by Linhart (1960). Fligner and Verducci (1987) also discussed measures of intergroup concordance. Feigin and Alvo (1986) developed a general approach to measure diversity and concordance. See also Mallows (1957), Hays (1960), Feigin and Cohen (1978), Hubert (1979), Cohen (1982), Snell (1983). Applications to consumer and marketing research are treated by Ryans (1976), Ryans and Srinavasan (1979), and Palachek and Kerin (1982).

Comparison of *n* rankings with a criterion ranking

13.3   Lyerly (1952) considered the problem of comparing the average correlation of *M* independent sets of rankings of *n* objects with a single dependent or criterion ranking of the same *n* objects, with no interest in the correlation within the *n* sets of ranks. He gave a formula to calculate the average inter-rho that does not require calculation of the individual rank correlations between the *mn* pairs of criterion–judgment ranks. The exact null and asymptotic distributions were given. Cureton (1958) gave a method of correcting the average inter-rho for ties. Hutchinson (1976) gave a related test and a table of critical values.

Jonckheere (1954b) and Page (1963) developed tests for agreement based on average Kendall tau values between a set of rankings of independent judges and an external or hypothetical ranking.

Palachek and Schucany (1983) extended this to the situation where the agreement between the external and internal orderings is substantial relative to the strength of the agreement between internal orderings.

These methods are special cases of the general problem of two-group concordance where one of the groups is regarded as the criterion ranking or external ranking.

Uses of rank correlation in linear regression

13.4   Consider a random sample of *n* pairs (X_i, Y_i) and the simple linear regression model with

$$Y_i = \alpha + \beta X_i + e_i$$

where α is the intercept parameter, β is the slope parameter, and the e_i are independent and identically distributed random variables with an unspecified continuous distribution. We assume without loss of generality that

$$X_1 < X_2 < \cdots < X_n.$$

A natural point estimate $\hat{\beta}$ of β is the median of the $n(n-1)/2$ sample estimates of slope

$$S_{ji} = \frac{Y_j - Y_i}{X_j - X_i}, \qquad 1 \le i < j \le n$$

and a natural point estimate of α is $\hat{\alpha}$, the median of

$$Y_i - \hat{\beta} X_i.$$

A test of the null hypothesis $\beta = \beta_0$ is provided by $S = P - Q$, where

$$P = \text{number of positive } S_{ji} = \text{number of concordant pairs}$$

$$Q = \text{number of negative } S_{ji} = \text{number of discordant pairs},$$

among the $(X_i, Y_i - \alpha - \beta_0 X_i)$ pairs, or equivalently among the $(X_i, Y_i - \beta_0 X_i)$. This test is then equivalent to the significance tests for Kendall's tau described in Section 4.7 and using Appendix Table 1 or the normal approximation described in Section 4.8. Corresponding methods for finding confidence interval estimates of β are described by Noether (1985).

This test for β based on the Kendall tau is due to Theil (1950). Sen (1968) discussed its asymptotic normality and efficiency and gave a procedure for handling ties. See Randles (1988) for additional references. A similar test can be based on the sum of squared differences of the ranks of the X_is and the corresponding values $Y_i - \beta_0 X_i$, or equivalently on the Spearman rank correlation coefficient between these pairs. Ghosh (1975) gave a sequential procedure based on rho for bounded length confidence intervals in simple regression.

Some references on efficiency were given in Chapter 5. Other important references are Konijn (1961), Adichie (1967a, 1967b).

Modified versions of these estimation procedures and tests based on tau and rho were proposed by Markowski (1984), Friedman and Rafsky (1983), Sievers (1978), Scholz (1978), Bhattacharyya (1968), and Sen (1968). Tritchler (1988) extended the test and confidence interval procedures based on tau to the case of multiple linear regression. Salama and Quade (1982) used a weighted measure of rank correlation to compare two multiple regressions. David and Fix (1961) discussed rank correlation and regression in a non-normal surface.

Lancaster and Quade (1984) developed a simultaneous test for both α and β that is based on a combination of the Kendall tau coefficient between the residuals $Y_i - \beta_0 X_i - \alpha_0$ and the constants X_i, and the sign test on the residuals. They gave a table of the null distribution, discussed approximations, and developed a joint confidence region for α and β based on the Daniels (1954) concept.

Bhattacharyya (1984) gave a comprehesive summary of non-parametric tests of randomness against trend or serial correlation alternatives with many references.

Power and efficiency of rank correlation methods

13.5 A great deal of work has been and is still being done on the power and efficiency of tests and estimation procedures based on rank correlation methods. This research is important and should be continued. However, a test or estimation procedure based on ranks is primarily used because of its generality and its lack of dependence upon specific distribution assumptions.

Some references for power and efficiency studies have already been made in the Notes and References sections of previous chapters. These are by no means complete, but will have to suffice for this book. Additional relevant references are given in Kendall and Stuart (1979), Van Dantzig and Hemelrijk (1954), and Savage (1952, 1954).

Appendix Table 1

Probability function of S and t (Kendall)

Each table entry for $n \leq 10$ is the probability that S (for Kendall's tau) is greater than or equal to a specified positive value S_0. The same entry is the probability that S is less than or equal to the corresponding negative S_0.

Each entry for $11 \leq n \leq 30$ is the significance point of a positive t (Kendall's tau) for a test at the labelled one-tailed level of significance, or equivalently, the smallest value of a positive t for which the right-tail probability for a one-sided test is less than or equal to the labelled values. Corresponding significance points for a negative t are found by symmetry. These values are from Dunstan, Nix and Reynolds (1979) with permission of the authors and the editor of RND publications.

Repeated zeros are indicated by powers; e.g. $0.0^3 47$ stands for 0.00047.

	Values of n					Values of n		
S	4	5	8	9	S	6	7	10
0	0.625	0.592	0.548	0.540	1	0.500	0.500	0.500
2	0.375	0.408	0.452	0.460	3	0.360	0.386	0.431
4	0.167	0.242	0.360	0.381	5	0.235	0.281	0.364
6	0.042	0.117	0.274	0.306	7	0.136	0.191	0.300
8		0.042	0.199	0.238	9	0.068	0.119	0.242
10		$0.0^2 83$	0.138	0.179	11	0.028	0.068	0.190
12			0.089	0.130	13	$0.0^2 83$	0.035	0.146
14			0.054	0.090	15	$0.0^2 14$	0.015	0.108
16			0.031	0.060	17		$0.0^2 54$	0.078
18			0.016	0.038	19		$0.0^2 14$	0.054
20			$0.0^2 71$	0.022	21		$0.0^3 20$	0.036
22			$0.0^2 28$	0.012	23			0.023
24			$0.0^3 87$	$0.0^2 63$	25			0.014
26			$0.0^3 19$	$0.0^2 29$	27			$0.0^2 83$
28			$0.0^4 25$	$0.0^2 12$	29			$0.0^2 46$
30				$0.0^3 43$	31			$0.0^2 23$
32				$0.0^3 12$	33			$0.0^2 11$
34				$0.0^4 25$	35			$0.0^3 47$
36				$0.0^5 28$	37			$0.0^3 18$
					39			$0.0^4 58$
					41			$0.0^4 15$
					43			$0.0^5 28$
					45			$0.0^6 28$

Appendix Table 1—*continued*

| | | Significance points of t | | | |
| | | One-tailed level of significance | | | |
n	0.10	0.05	0.025	0.01	0.005
11	0.345	0.418	0.491	0.564	0.600
12	0.303	0.394	0.455	0.545	0.576
13	0.308	0.359	0.436	0.513	0.564
14	0.275	0.363	0.407	0.473	0.516
15	0.276	0.333	0.390	0.467	0.505
16	0.250	0.317	0.383	0.433	0.483
17	0.250	0.309	0.368	0.426	0.471
18	0.242	0.294	0.346	0.412	0.451
19	0.228	0.287	0.333	0.392	0.439
20	0.221	0.274	0.326	0.379	0.421
21	0.210	0.267	0.314	0.371	0.410
22	0.203	0.264	0.307	0.359	0.394
23	0.202	0.257	0.296	0.352	0.391
24	0.196	0.246	0.290	0.341	0.377
25	0.193	0.240	0.287	0.333	0.367
26	0.188	0.237	0.280	0.329	0.360
27	0.180	0.231	0.271	0.322	0.356
28	0.180	0.228	0.265	0.312	0.344
29	0.172	0.222	0.261	0.310	0.340
30	0.172	0.218	0.255	0.301	0.333

Appendix Table 2

Probability function of $\sum d^2$ (for r_s)

This table lists for $n \leq 16$ only those values of $\sum d^2 > n(n^2 - 1)/6$, which are those with left-tail P-values ≤ 0.5. To find right-tail P-values for $\sum d^2 < n(n^2 - 1)/6$, enter the table with $n(n^2 - 1)/3 - \sum d^2$. For example, with $n = 4$, $\sum d^2 = 8$, enter the table with $20 - 8 = 12$ and read out a right-tail P-value of 0.458. Decimal points are omitted. For $n \geq 17$ this table gives significance points for positive values of r_s. Corresponding significance points for negative r_s are the same by symmetry.

Values for $n = 14$ through $n = 16$ are from Franklin (1988c) with permission of the author and the *Journal of Statistical Computation and Simulation*. Values for $n \geq 17$ are from Ramsay (1989) with permission of the author and editor of the *Journal of Educational Statistics*.

$n = 4$		$n = 5$		$n = 6$		$n = 7$		$n = 8$	
$\sum d^2$	P	$\sum d^2$	P	$\sum d^2$	P	$\sum d^2$	P	$\sum d^2$	P
12	458	22	475	36	500	58	482	86	488
14	375	24	392	38	460	60	453	88	467
16	208	26	342	40	401	62	420	90	441
18	167	28	258	42	357	64	391	92	420
20	042	30	225	44	329	66	357	94	397
		32	175	46	282	68	331	96	376
		34	117	48	249	70	297	98	352
		36	067	50	210	72	278	100	332
		38	042	52	178	74	249	102	310
		40	0^283	54	149	76	222	104	291
				56	121	78	198	106	268
				58	088	80	177	108	250
				60	068	82	151	110	231
				62	051	84	133	112	214
				64	029	86	118	114	195
				66	017	88	100	116	180
				68	0^283	90	083	118	163
				70	0^214	92	069	120	150
						94	055	122	134
						96	044	124	122
						98	033	126	108
						100	024	128	098
						102	017	130	085
						104	012	132	076
						106	0^262	134	066
						108	0^234	136	057
						110	0^214	138	048
						112	0^320	140	042
								142	035
								144	029
								146	023

Appendix Table 2—*continued*

n = 8		n = 9		n = 9		n = 10		n = 10	
$\sum d^2$	P	$\sum d^2$	P	$\sum d^2$	P	$\sum d^2$	P	$\sum d^2$	P
148	018	122	491	220	0^241	166	500	264	037
150	014	124	474	222	0^230	168	486	266	033
152	011	126	456	224	0^223	170	473	268	030
154	0^277	128	440	226	0^215	172	459	270	027
156	0^254	130	422	228	0^210	174	446	272	024
158	0^236	132	405	230	0^366	176	433	274	022
160	0^223	134	388	232	0^337	178	419	276	019
162	0^211	136	372	234	0^318	180	406	278	017
164	0^357	138	354	236	0^483	182	393	280	015
166	0^320	140	339	238	0^425	184	379	282	013
168	0^425	142	322	240	0^528	186	367	284	012
		144	307			188	354	286	010
		146	290			190	341	288	0^287
		148	276			192	328	290	0^275
		150	260			194	316	292	0^263
		152	247			196	304	294	0^253
		154	231			198	292	296	0^244
		156	218			200	280	298	0^236
		158	205			202	268	300	0^229
		160	193			204	257	302	0^224
		162	179			206	246	304	0^219
		164	168			208	235	306	0^214
		166	156			210	224	308	0^211
		168	146			212	214	310	0^380
		170	135			214	203	312	0^357
		172	125			216	193	314	0^340
		174	115			218	184	316	0^327
		176	106			220	174	318	0^317
		178	097			222	165	320	0^310
		180	089			224	156	322	0^454
		182	081			226	148	324	0^425
		184	074			228	139	326	0^410
		186	066			230	132	328	0^528
		188	060			232	124	330	0^628
		190	054			234	116		
		192	048			236	109		
		194	043			238	102		
		196	038			240	096		
		198	033			242	089		
		200	029			244	083		
		202	025			246	077		
		204	022			248	072		
		206	018			250	067		
		208	016			252	062		
		210	013			254	057		
		212	011			256	052		
		214	0^286			258	048		
		216	0^269			260	044		
		218	0^254			262	040		

Appendix Table 2—*continued*

$n = 11$		$n = 11$		$n = 11$		$n = 12$		$n = 12$	
$\sum d^2$	P	$\sum d^2$	P	$\sum d^2$	P	$\sum d^2$	P	$\sum d^2$	P
222	495	320	082	418	$0^3 19$	288	496	386	133
224	484	322	077	420	$0^3 14$	290	487	388	128
226	473	324	073	422	$0^4 95$	292	478	390	123
228	462	326	069	424	$0^4 64$	294	469	392	118
230	452	328	065	426	$0^4 41$	296	460	394	114
232	441	330	061	428	$0^4 24$	298	452	396	109
234	430	332	057	430	$0^4 14$	300	443	398	105
236	419	334	054	432	$0^5 69$	302	435	400	100
238	409	336	050	434	$0^5 30$	304	426	402	096
240	398	338	047	436	$0^5 12$	306	417	404	092
242	388	340	044	438	$0^6 28$	308	409	406	088
244	377	342	041	440	$0^7 25$	310	400	408	084
246	367	344	038			312	391	410	081
248	357	346	035			314	383	412	077
250	347	348	033			316	375	414	074
252	337	350	030			318	366	416	070
254	327	352	028			320	358	418	067
256	317	354	026			322	350	420	064
258	307	356	024			324	342	422	061
260	298	358	022			326	334	424	058
262	288	360	020			328	325	426	055
264	279	362	018			330	318	428	052
266	270	364	017			332	310	430	049
268	260	366	015			334	302	432	047
270	252	368	014			336	294	434	044
272	243	370	013			338	287	436	042
274	234	372	011			340	279	438	040
276	226	374	010			342	272	440	037
278	217	376	$0^2 91$			344	264	442	035
280	209	378	$0^2 81$			346	257	444	033
282	201	380	$0^2 72$			348	250	446	031
284	193	382	$0^2 64$			350	243	448	030
286	186	384	$0^2 56$			352	235	450	028
288	178	386	$0^2 49$			354	229	452	026
290	171	388	$0^2 43$			356	222	454	024
292	163	390	$0^2 37$			358	215	456	023
294	157	392	$0^2 32$			360	208	458	021
296	150	394	$0^2 27$			362	202	460	020
298	143	396	$0^2 23$			364	196	462	019
300	137	398	$0^2 20$			366	189	464	017
302	130	400	$0^2 16$			368	183	466	016
304	124	402	$0^2 14$			370	177	468	015
306	118	404	$0^2 11$			372	171	470	014
308	112	406	$0^3 91$			374	166	472	013
310	107	408	$0^3 73$			376	160	474	012
312	102	410	$0^3 58$			378	154	476	011
314	096	412	$0^3 45$			380	149	478	010
316	091	414	$0^3 35$			382	143	480	$0^2 93$
318	087	416	$0^3 26$			384	138	482	$0^2 85$

Appendix Table 2—*continued*

n = 12		n = 13		n = 13		n = 13		n = 13	
$\sum d^2$	P	$\sum d^2$	P	$\sum d^2$	P	$\sum d^2$	P	$\sum d^2$	P
484	$0^2 78$	366	496	464	182	562	029	660	$0^3 60$
486	$0^2 71$	368	489	466	177	564	028	662	$0^3 53$
488	$0^2 65$	370	482	468	172	566	026	664	$0^3 46$
490	$0^2 59$	372	475	470	167	568	025	666	$0^3 40$
492	$0^2 53$	374	468	472	162	570	024	668	$0^3 34$
494	$0^2 48$	376	460	474	158	572	022	670	$0^3 29$
496	$0^2 43$	378	453	476	153	574	021	672	$0^3 25$
498	$0^2 39$	380	446	478	149	576	020	674	$0^3 21$
500	$0^2 35$	382	439	480	144	578	019	676	$0^3 18$
502	$0^2 31$	384	432	482	140	580	018	678	$0^3 15$
504	$0^2 28$	386	425	484	136	582	017	680	$0^3 13$
506	$0^2 25$	388	417	486	132	584	016	682	$0^3 10$
508	$0^2 22$	390	410	488	128	586	015	684	$0^4 86$
510	$0^2 19$	392	403	490	124	588	014	686	$0^4 70$
512	$0^2 17$	394	396	492	120	590	013	688	$0^4 56$
514	$0^2 15$	396	389	494	116	592	013	690	$0^4 45$
516	$0^2 13$	398	382	496	112	594	012	692	$0^4 35$
518	$0^2 11$	400	375	498	108	596	011	694	$0^4 27$
520	$0^3 93$	402	369	500	105	598	010	696	$0^4 21$
522	$0^3 80$	404	362	502	101	600	$0^2 97$	698	$0^4 16$
524	$0^3 67$	406	355	504	098	602	$0^2 91$	700	$0^4 12$
526	$0^3 57$	408	348	506	094	604	$0^2 85$	702	$0^5 85$
528	$0^3 47$	410	341	508	091	606	$0^2 79$	704	$0^5 61$
530	$0^3 39$	412	335	510	088	608	$0^2 74$	706	$0^5 42$
532	$0^3 32$	414	328	512	085	610	$0^2 68$	708	$0^5 28$
534	$0^3 26$	416	321	514	082	612	$0^2 64$	710	$0^5 18$
536	$0^3 21$	418	315	516	079	614	$0^2 59$	712	$0^5 11$
538	$0^3 17$	420	308	518	076	616	$0^2 55$	714	$0^6 67$
540	$0^3 13$	422	302	520	073	618	$0^2 51$	716	$0^6 37$
542	$0^3 10$	424	296	522	070	620	$0^2 47$	718	$0^6 19$
544	$0^4 77$	426	289	524	068	622	$0^2 43$	720	$0^7 85$
546	$0^4 58$	428	283	526	065	624	$0^2 40$	722	$0^7 34$
548	$0^4 42$	430	277	528	062	626	$0^2 36$	724	$0^7 11$
550	$0^4 30$	432	271	530	060	628	$0^2 33$	726	$0^8 21$
552	$0^4 21$	434	265	532	057	630	$0^2 31$	728	$0^9 16$
554	$0^4 14$	436	259	534	055	632	$0^2 28$		
556	$0^5 90$	438	253	536	053	634	$0^2 25$		
558	$0^5 55$	440	247	538	051	636	$0^2 23$		
560	$0^5 32$	442	241	540	049	638	$0^2 21$		
562	$0^5 17$	444	235	542	046	640	$0^2 19$		
564	$0^6 80$	446	230	544	044	642	$0^2 17$		
566	$0^6 34$	448	224	546	043	644	$0^2 15$		
568	$0^6 12$	450	218	548	041	646	$0^2 14$		
570	$0^7 25$	452	213	550	039	648	$0^2 12$		
572	$0^8 21$	454	208	552	037	650	$0^2 11$		
		456	202	554	035	652	$0^3 99$		
		458	197	556	034	654	$0^3 88$		
		460	192	558	032	656	$0^3 78$		
		462	187	560	031	658	$0^3 68$		

Appendix Table 2—*continued*

$\sum d^2$	P	$\sum d^2$	P	$\sum d^2$	P	$\sum d^2$	P	$\sum d^2$	P
	$n = 14$		$n = 14$		$n = 14$		$n = 14$		$n = 14$
456	500	554	227	652	062	750	$0^2$72	846	$0^4$87
458	494	556	222	654	060	752	$0^2$68	848	$0^4$74
460	488	558	218	656	058	754	$0^2$64	850	$0^4$63
462	482	560	213	658	056	756	$0^2$60	852	$0^4$54
464	476	562	209	660	054	758	$0^2$57	854	$0^4$45
466	470	564	204	662	052	760	$0^2$53	856	$0^4$38
468	464	566	200	664	050	762	$0^2$50	858	$0^4$31
470	458	568	196	666	049	764	$0^2$47	860	$0^4$26
472	452	570	191	668	047	766	$0^2$44	862	$0^4$21
474	446	572	187	670	045	768	$0^2$41	864	$0^4$17
476	440	574	183	672	044	770	$0^2$38	866	$0^4$14
478	434	576	179	674	042	772	$0^2$36	868	$0^4$11
480	428	578	175	676	040	774	$0^2$34	870	$0^5$88
482	422	580	171	678	039	776	$0^2$31	872	$0^5$69
484	416	582	167	680	038	778	$0^2$29	874	$0^5$53
486	410	584	163	682	036	780	$0^2$27	876	$0^5$40
488	404	586	159	684	035	782	$0^2$25	878	$0^5$30
490	398	588	155	686	033	784	$0^2$23	880	$0^5$22
492	392	590	151	688	032	786	$0^2$22	882	$0^5$16
494	387	592	148	690	031	788	$0^2$20	884	$0^5$12
496	381	594	144	692	030	790	$0^2$18	886	$0^6$80
498	375	596	140	694	028	792	$0^2$17	888	$0^6$53
500	369	598	137	696	027	794	$0^2$16	890	$0^6$35
502	364	600	133	698	026	796	$0^2$14	892	$0^6$22
504	358	602	130	700	025	798	$0^2$13	894	$0^6$13
506	352	604	127	702	024	800	$0^2$12	896	$0^7$74
508	346	606	123	704	023	802	$0^2$11	898	$0^7$40
510	341	608	120	706	022	804	$0^2$10	900	$0^7$19
512	335	610	117	708	021	806	$0^3$93	902	$0^8$82
514	330	612	114	710	020	808	$0^3$85	904	$0^8$31
516	324	614	111	712	019	810	$0^3$77	906	$0^9$92
518	319	616	107	714	018	812	$0^3$70	908	$0^9$16
520	313	618	104	716	018	814	$0^3$63	910	0^{10}11
522	308	620	102	718	017	816	$0^3$57		
524	303	622	099	720	016	818	$0^3$51		
526	297	624	096	722	015	820	$0^3$46		
528	292	626	093	724	014	822	$0^3$41		
530	287	628	090	726	014	824	$0^3$37		
532	281	630	088	728	013	826	$0^3$33		
534	276	632	085	730	012	828	$0^3$29		
536	271	634	083	732	012	830	$0^3$26		
538	266	636	080	734	011	832	$0^3$23		
540	261	638	078	736	011	834	$0^3$20		
542	256	640	075	738	010	836	$0^3$18		
544	251	642	073	740	$0^2$95	838	$0^3$15		
546	246	644	071	742	$0^2$90	840	$0^3$13		
548	241	646	068	744	$0^2$85	842	$0^3$12		
550	237	648	066	746	$0^2$81	844	$0^3$10		
552	232	650	064	748	$0^2$76	846	$0^4$87		

Appendix Table 2—*continued*

$n = 15$		$n = 15$		$n = 15$		$n = 15$		$n = 15$	
Σd^2	P	Σd^2	P	Σd^2	P	Σd^2	P	Σd^2	P
562	497	660	262	758	098	856	023	954	$0^2 22$
564	492	662	258	760	096	858	022	956	$0^2 21$
566	487	664	253	762	094	860	021	958	$0^2 20$
568	482	666	249	764	091	862	020	960	$0^2 19$
570	477	668	245	766	089	864	020	962	$0^2 18$
572	472	670	241	768	087	866	019	964	$0^2 16$
574	467	672	237	770	085	868	018	966	$0^2 15$
576	462	674	233	772	082	870	017	968	$0^2 14$
578	457	676	229	774	080	872	017	970	$0^2 13$
580	452	678	225	776	078	874	016	972	$0^2 12$
582	446	680	221	778	076	876	016	974	$0^2 12$
584	441	682	217	780	074	878	015	976	$0^2 11$
586	436	684	213	782	072	880	014	978	$0^2 10$
588	431	686	209	784	070	882	014	980	$0^3 94$
590	426	688	206	786	068	884	013	982	$0^3 87$
592	421	690	202	788	067	886	013	984	$0^3 80$
594	416	692	198	790	065	888	012	986	$0^3 74$
596	411	694	195	792	063	890	012	988	$0^3 69$
598	406	696	191	794	061	892	011	990	$0^3 64$
600	401	698	187	796	060	894	011	992	$0^3 59$
602	396	700	184	798	058	896	010	994	$0^3 54$
604	391	702	180	800	056	898	$0^2 97$	996	$0^3 50$
606	386	704	177	802	055	900	$0^2 93$	998	$0^3 46$
608	382	706	173	804	053	902	$0^2 89$	1000	$0^3 42$
610	377	708	170	806	052	904	$0^2 85$	1002	$0^3 38$
612	372	710	167	808	050	906	$0^2 81$	1004	$0^3 35$
614	367	712	163	810	049	908	$0^2 77$	1006	$0^3 32$
616	362	714	160	812	047	910	$0^2 74$	1008	$0^3 29$
618	357	716	157	814	046	912	$0^2 70$	1010	$0^3 26$
620	352	718	154	816	044	914	$0^2 67$	1012	$0^3 24$
622	348	720	151	818	043	916	$0^2 64$	1014	$0^3 22$
624	343	722	147	820	042	918	$0^2 61$	1016	$0^3 20$
626	338	724	144	822	040	920	$0^2 58$	1018	$0^3 18$
628	333	726	141	824	039	922	$0^2 55$	1020	$0^3 16$
630	329	728	138	826	038	924	$0^2 52$	1022	$0^3 14$
632	324	730	135	828	037	926	$0^2 50$	1024	$0^3 13$
634	319	732	132	830	036	928	$0^2 47$	1026	$0^3 12$
636	315	734	130	832	034	930	$0^2 45$	1028	$0^3 10$
638	310	736	127	834	033	932	$0^2 42$	1030	$0^4 91$
640	306	738	124	836	032	934	$0^2 40$	1032	$0^4 81$
642	301	740	121	838	031	936	$0^2 38$	1034	$0^4 72$
644	297	742	118	840	030	938	$0^2 36$	1036	$0^4 63$
646	292	744	116	842	029	940	$0^2 34$	1038	$0^4 56$
648	288	746	113	844	028	942	$0^2 32$	1040	$0^4 49$
650	283	748	111	846	027	944	$0^2 30$	1042	$0^4 42$
652	279	750	108	848	026	946	$0^2 29$	1044	$0^4 37$
654	275	752	106	850	025	948	$0^2 27$	1046	$0^4 32$
656	270	754	103	852	024	950	$0^2 26$	1048	$0^4 28$
658	266	756	101	854	024	952	$0^2 24$	1050	$0^4 24$

Appendix Table 2—*continued*

$n = 15$		$n = 16$		$n = 16$		$n = 16$		$n = 16$	
$\sum d^2$	P	$\sum d^2$	P	$\sum d^2$	P	$\sum d^2$	P	$\sum d^2$	P
1052	$0^4 20$	682	498	780	293	878	137	976	047
1054	$0^4 18$	684	493	782	289	880	134	978	046
1056	$0^4 15$	686	489	784	286	882	132	980	044
1058	$0^4 12$	688	485	786	282	884	129	982	043
1060	$0^4 10$	690	480	788	278	886	127	984	042
1062	$0^5 88$	692	476	790	274	888	124	986	041
1064	$0^5 73$	694	472	792	271	890	122	988	040
1066	$0^5 60$	696	467	794	267	892	120	990	039
1068	$0^5 49$	698	463	796	263	894	117	992	038
1070	$0^5 40$	700	459	798	260	896	115	994	037
1072	$0^5 32$	702	454	800	256	898	113	996	036
1074	$0^5 26$	704	450	802	252	900	111	998	035
1076	$0^5 21$	706	446	804	249	902	108	1000	034
1078	$0^5 16$	708	441	806	245	904	106	1002	033
1080	$0^5 12$	710	437	808	242	906	104	1004	032
1082	$0^6 96$	712	433	810	239	908	102	1006	031
1084	$0^6 73$	714	428	812	235	910	100	1008	030
1086	$0^6 54$	716	424	814	232	912	098	1010	029
1088	$0^6 40$	718	420	816	228	914	096	1012	029
1090	$0^6 29$	720	415	818	225	916	094	1014	028
1092	$0^6 20$	722	411	820	221	918	092	1016	027
1094	$0^6 14$	724	407	822	218	920	090	1018	026
1096	$0^7 95$	726	403	824	215	922	088	1020	025
1098	$0^7 62$	728	398	826	212	924	086	1022	025
1100	$0^7 39$	730	394	828	208	926	084	1024	024
1102	$0^7 24$	732	390	830	205	928	083	1026	023
1104	$0^7 14$	734	386	832	202	930	081	1028	022
1106	$0^8 75$	736	381	834	199	932	079	1030	022
1108	$0^8 39$	738	377	836	196	934	077	1032	021
1110	$0^8 18$	740	373	838	193	936	076	1034	020
1112	$0^9 72$	742	369	840	190	938	074	1036	020
1114	$0^9 26$	744	365	842	186	940	072	1038	019
1116	$0^{10} 71$	746	361	844	183	942	071	1040	019
1118	$0^{10} 11$	748	357	846	180	944	069	1042	018
1120	$0^{12} 76$	750	352	848	178	946	068	1044	017
		752	348	850	175	948	066	1046	017
		754	344	852	172	950	064	1048	016
		756	340	854	169	952	063	1050	016
		758	336	856	166	954	062	1052	015
		760	332	858	163	956	060	1054	015
		762	328	860	160	958	059	1056	014
		764	324	862	158	960	057	1058	014
		766	320	864	155	962	056	1060	013
		768	316	866	152	964	054	1062	013
		770	312	868	149	966	053	1064	012
		772	309	870	147	968	052	1066	012
		774	305	872	144	970	050	1068	012
		776	301	874	142	972	049	1070	011
		778	297	876	139	974	048	1072	011

Appendix Table 2—*concluded*

$n = 16$		$n = 16$		$n = 16$		$n = 16$		$n = 16$	
$\sum d^2$	P	$\sum d^2$	P	$\sum d^2$	P	$\sum d^2$	P	$\sum d^2$	P
1074	010	1132	0^231	1190	0^361	1248	0^458	1306	0^688
1076	010	1134	0^230	1192	0^358	1250	0^452	1308	0^671
1078	010	1136	0^228	1194	0^354	1252	0^447	1310	0^657
1080	0^293	1138	0^227	1196	0^350	1254	0^442	1312	0^645
1082	0^289	1140	0^226	1198	0^347	1256	0^438	1314	0^636
1084	0^286	1142	0^224	1200	0^344	1258	0^434	1316	0^628
1086	0^283	1144	0^223	1202	0^341	1260	0^430	1318	0^621
1088	0^280	1146	0^222	1204	0^338	1262	0^427	1320	0^616
1090	0^276	1148	0^221	1206	0^335	1264	0^424	1322	0^612
1092	0^274	1150	0^220	1208	0^333	1266	0^421	1324	0^791
1094	0^271	1152	0^219	1210	0^330	1268	0^419	1326	0^767
1096	0^268	1154	0^218	1212	0^328	1270	0^416	1328	0^748
1098	0^265	1156	0^217	1214	0^326	1272	0^414	1330	0^734
1100	0^263	1158	0^216	1216	0^324	1274	0^413	1332	0^723
1102	0^260	1160	0^215	1218	0^322	1276	0^411	1334	0^716
1104	0^258	1162	0^214	1220	0^320	1278	0^596	1336	0^710
1106	0^255	1164	0^214	1222	0^319	1280	0^583	1338	0^866
1108	0^253	1166	0^213	1224	0^317	1282	0^572	1340	0^840
1110	0^251	1168	0^212	1226	0^316	1284	0^562	1342	0^824
1112	0^249	1170	0^211	1228	0^315	1286	0^553	1344	0^813
1114	0^246	1172	0^211	1230	0^314	1288	0^545	1346	0^970
1116	0^244	1174	0^210	1232	0^312	1290	0^539	1348	0^935
1118	0^243	1176	0^210	1234	0^311	1292	0^533	1350	0^915
1120	0^241	1178	0^390	1236	0^310	1294	0^528	1352	$0^{10}59$
1122	0^239	1180	0^384	1238	0^494	1296	0^523	1354	$0^{10}20$
1124	0^237	1182	0^379	1240	0^485	1298	0^519	1356	$0^{11}51$
1126	0^236	1184	0^374	1242	0^478	1300	0^516	1358	$0^{12}76$
1128	0^234	1186	0^370	1244	0^470	1302	0^513	1360	$0^{13}48$
1130	0^232	1188	0^366	1246	0^464	1304	0^511		

Significance points of r_s

n	0.10	0.05	0.025	0.01	0.005	0.001
17	0.328	0.414	0.488	0.566	0.618	0.711
18	0.317	0.401	0.472	0.550	0.600	0.692
19	0.309	0.391	0.460	0.535	0.584	0.675
20	0.299	0.380	0.447	0.522	0.570	0.662
21	0.292	0.370	0.436	0.509	0.556	0.647
22	0.284	0.361	0.425	0.497	0.544	0.633
23	0.278	0.353	0.416	0.486	0.532	0.621
24	0.271	0.344	0.407	0.476	0.521	0.609
25	0.265	0.337	0.398	0.466	0.511	0.597
26	0.259	0.331	0.390	0.457	0.501	0.586
27	0.255	0.324	0.383	0.449	0.492	0.576
28	0.250	0.318	0.375	0.441	0.483	0.567
29	0.245	0.312	0.368	0.433	0.475	0.558
30	0.240	0.306	0.362	0.425	0.467	0.549
31	0.236	0.301	0.356	0.419	0.459	0.540
32	0.232	0.296	0.350	0.412	0.452	0.532
33	0.229	0.291	0.345	0.405	0.446	0.525
34	0.225	0.287	0.340	0.400	0.439	0.517
35	0.222	0.283	0.335	0.394	0.433	0.510

Appendix Table 3

Probability function of the standard normal distribution

This table shows the area under the standard normal distribution lying to the left of a positive normal deviate z. This same value is the area lying to the right of a corresponding negative deviate z. For example, the area to the left of z = 1.86 (1.5 + 0.36) is 0.9686 and the area to the right of z = −1.86 is 0.9686. The area to the left of z = −1.86 is 1 − 0.9686 = 0.0314.

Decimal points in the body of the table are omitted. Repeated nines are indicated by powers; e.g. 9^371 stands for 0.99971.

Deviate	0. +	0.5 +	1.0 +	1.5 +	2.0 +	2.5 +	3.0 +	3.5 +
0.00	5000	6915	8413	9332	9772	9^2379	9^2865	9^377
0.01	5040	6950	8438	9345	9778	9^2396	9^2869	9^378
0.02	5080	6985	8461	9357	9783	9^2413	9^2874	9^378
0.03	5120	7019	8485	9370	9788	9^2430	9^2878	9^379
0.04	5160	7054	8508	9382	9793	9^2446	9^2882	9^380
0.05	5199	7088	8531	9394	9798	9^2461	9^2886	9^381
0.06	5239	7123	8554	9406	9803	9^2477	9^2889	9^381
0.07	5279	7157	8577	9418	9808	9^2492	9^2893	9^382
0.08	5319	7190	8599	9429	9812	9^2506	9^2897	9^383
0.09	5359	7224	8621	9441	9817	9^2520	9^2900	9^383
0.10	5398	7257	8643	9452	9821	9^2534	9^303	9^384
0.11	5438	7291	8665	9463	9826	9^2547	9^306	9^385
0.12	5478	7324	8686	9474	9830	9^2560	9^310	9^385
0.13	5517	7357	8708	9484	9834	9^2573	9^313	9^386
0.14	5557	7389	8729	9495	9838	9^2585	9^316	9^386
0.15	5596	7422	8749	9505	9842	9^2598	9^318	9^387
0.16	5636	7454	8770	9515	9846	9^2609	9^321	9^387
0.17	5675	7486	8790	9525	9850	9^2621	9^324	9^388
0.18	5714	7517	8810	9535	9854	9^2632	9^326	9^388
0.19	5753	7549	8830	9545	9857	9^2643	9^329	9^389
0.20	5793	7580	8849	9554	9861	9^2653	9^331	9^389
0.21	5832	7611	8869	9564	9864	9^2664	9^334	9^390
0.22	5871	7642	8888	9573	9868	9^2674	9^336	9^390
0.23	5910	7673	8907	9582	9871	9^2683	9^338	9^404
0.24	5948	7704	8925	9591	9875	9^2693	9^340	9^408
0.25	5987	7738	8944	9599	9878	9^2702	9^342	9^412
0.26	6026	7764	8962	9608	9881	9^2711	9^344	9^415
0.27	6064	7794	8980	9616	9884	9^2720	9^346	9^418
0.28	6103	7823	8997	9625	9887	9^2728	9^348	9^422
0.29	6141	7852	9015	9633	9890	9^2736	9^350	9^425
0.30	6179	7881	9032	9641	9893	9^2744	9^352	9^428
0.31	6217	7910	9049	9649	9896	9^2752	9^353	9^431
0.32	6255	7939	9066	9656	9898	9^2760	9^355	9^433
0.33	6293	7967	9082	9664	9901	9^2767	9^357	9^436
0.34	6331	7995	9099	9671	9904	9^2774	9^358	9^439
0.35	6368	8023	9115	9678	9906	9^2781	9^360	9^441
0.36	6406	8051	9131	9686	9909	9^2788	9^361	9^443
0.37	6443	8078	9147	9693	9911	9^2795	9^362	9^446
0.38	6480	8106	9162	9699	9913	9^2801	9^364	9^448
0.39	6517	8133	9177	9706	9916	9^2807	9^365	9^450
0.40	6554	8159	9192	9713	9918	9^2813	9^366	9^452
0.41	6591	8186	9207	9719	9920	9^2819	9^368	9^454
0.42	6628	8212	9222	9726	9922	9^2825	9^369	9^456
0.43	6664	8238	9236	9732	9925	9^2831	9^370	9^458
0.44	6700	8264	9251	9738	9927	9^2836	9^371	9^459
0.45	6736	8289	9265	9744	9929	9^2841	9^372	9^461
0.46	6772	8315	9279	9750	9931	9^2846	9^373	9^463
0.47	6808	8340	9292	9756	9932	9^2851	9^374	9^464
0.48	6844	8365	9306	9761	9934	9^2856	9^375	9^466
0.49	6879	8389	9319	9767	9936	9^2861	9^376	9^467

Appendix Table 4

Random rankings of 20 (random permutations of the first twenty natural numbers)

Random rankings of fewer than 20 can be derived by merely omitting unwanted ranks.

4	19	15	10	13	3	11	18	1	8	7	2	14	17	12	9	6	20	5	16
16	14	2	17	8	11	12	13	4	3	7	10	5	15	1	6	20	18	19	9
6	5	10	12	17	7	9	2	1	15	19	11	13	14	16	3	20	4	18	8
18	9	10	19	6	12	4	1	20	11	13	16	15	3	2	5	14	17	8	7
11	12	8	9	1	15	13	2	14	17	3	19	6	7	16	4	5	20	18	10
1	15	18	4	8	20	19	9	11	3	17	10	13	14	6	7	5	2	12	16
18	2	16	12	8	5	11	17	4	10	3	9	7	13	1	14	6	19	20	15
2	3	18	19	11	1	7	12	20	8	15	5	14	4	13	6	9	17	16	10
15	16	12	13	17	19	1	2	20	7	4	11	10	9	6	8	5	18	14	3
8	18	3	4	10	7	16	17	9	6	11	15	19	2	13	12	14	1	20	5
8	6	13	1	10	11	15	20	7	14	3	18	2	12	16	17	4	19	5	9
10	5	17	2	15	9	16	3	14	1	6	12	19	13	20	4	11	8	7	18
16	1	18	11	2	14	13	15	20	7	9	3	19	17	5	12	6	10	4	8
12	7	5	20	15	19	13	10	18	1	8	16	9	2	6	11	4	3	17	14
20	1	19	5	9	6	11	10	7	3	2	18	4	14	8	17	15	12	13	16
9	6	17	15	10	12	18	13	1	19	4	5	14	3	2	16	7	11	20	8
4	2	3	19	11	10	7	16	13	18	15	1	12	17	9	8	6	5	20	14
9	18	19	14	10	12	4	16	5	2	6	17	1	8	20	11	13	7	3	15
13	12	8	7	18	4	16	10	17	19	5	1	20	3	14	11	2	6	9	15
6	1	4	5	8	9	10	18	15	20	13	7	12	2	3	19	17	16	11	14
8	20	4	5	15	10	11	6	16	18	19	9	7	14	17	1	12	3	2	13
19	16	3	4	7	6	12	1	13	2	10	17	8	20	11	9	18	5	15	14
19	18	13	8	15	9	10	2	17	14	16	7	5	1	20	4	11	3	6	12
2	11	13	5	7	18	20	6	16	14	1	3	15	8	19	9	12	17	4	10
2	12	16	10	17	13	20	11	3	4	8	14	15	19	7	9	5	6	18	1
13	11	4	3	7	5	20	15	1	12	9	6	14	17	16	2	18	19	10	8
15	1	16	7	3	14	8	11	2	13	10	6	18	5	9	12	19	4	17	20
3	20	1	11	17	16	4	14	6	2	9	19	15	12	10	8	5	13	7	18
18	16	3	2	12	19	4	20	14	11	8	10	7	17	1	5	15	13	9	6
3	4	12	5	10	14	15	1	7	2	13	19	17	18	6	11	20	16	9	8
13	20	11	14	3	9	4	5	19	17	18	1	16	15	8	10	6	7	12	2
17	16	13	18	7	2	9	15	10	6	11	4	12	19	14	3	20	5	1	8
13	7	3	4	8	6	15	2	18	16	5	19	9	12	10	20	11	1	14	17
2	19	17	7	18	20	15	5	9	11	3	16	13	6	12	1	4	10	14	8
4	6	7	10	20	18	3	5	11	12	2	1	14	9	19	13	15	17	8	16
19	3	15	12	10	17	16	1	8	7	2	6	14	9	11	18	4	5	20	13
18	14	9	2	8	6	7	1	3	10	5	17	20	4	13	15	11	12	16	19
18	5	17	1	4	7	11	20	9	10	15	3	19	14	16	13	6	12	8	2
5	1	10	6	19	7	15	11	2	4	3	9	20	18	14	17	13	16	8	12
19	17	1	16	10	9	8	18	13	15	5	2	7	20	14	12	11	4	3	6

Appendix Table 4—*continued*

13	6	14	4	9	17	19	7	12	8	18	16	11	1	3	5	20	10	2	15
10	16	9	6	12	17	4	19	11	1	20	13	14	18	15	7	2	3	8	5
14	8	7	16	18	6	17	19	11	20	4	13	2	10	1	15	9	3	5	12
16	2	8	14	4	10	19	11	7	6	12	17	5	20	13	15	1	9	3	18
14	5	9	8	17	10	6	12	19	13	3	20	2	18	4	7	1	11	15	16
12	18	15	4	1	13	5	17	9	7	14	20	19	6	10	11	2	3	8	16
8	17	5	3	4	12	10	14	9	18	15	6	13	2	20	16	19	1	7	11
11	19	8	9	16	10	13	17	15	4	5	20	14	2	1	3	7	18	6	12
14	1	5	2	3	15	11	9	18	19	7	10	16	13	6	8	4	17	12	20
9	17	5	20	2	7	4	19	18	15	1	10	8	16	14	3	11	13	12	6
10	14	17	19	20	1	13	18	5	11	7	8	16	3	9	12	15	6	2	4
20	5	9	14	1	8	4	12	13	3	15	7	2	19	10	6	17	16	11	18
5	15	1	17	2	11	9	10	13	18	19	16	3	14	4	20	12	8	6	7
2	13	14	3	9	19	15	8	20	17	18	4	11	10	6	1	5	7	12	16
18	12	14	10	1	13	9	3	8	7	5	17	16	2	4	20	15	19	6	11
3	16	11	6	12	13	15	5	7	20	19	18	4	10	9	14	8	2	17	1
2	11	4	16	13	20	14	18	6	15	8	10	1	3	19	7	5	9	17	12
14	12	20	10	15	2	9	3	16	13	6	5	7	11	8	1	17	19	4	18
12	10	18	5	4	20	17	16	13	15	9	7	14	2	6	11	1	19	8	3
1	18	5	16	12	7	19	15	14	2	10	9	13	3	20	4	17	11	6	8
10	8	14	18	16	3	2	4	9	19	15	5	11	12	13	17	6	1	20	7
18	4	1	9	12	14	15	19	2	6	3	10	17	13	7	16	8	20	11	5
1	15	18	9	13	16	7	8	5	11	3	4	12	20	14	10	19	6	2	17
13	20	11	10	6	12	18	9	15	17	19	1	7	16	8	2	3	5	14	4
6	16	9	7	10	2	8	20	13	1	11	15	17	3	19	4	14	12	18	5
20	7	10	5	15	19	16	1	13	12	11	18	4	3	6	2	14	17	9	8
17	12	19	4	2	6	1	15	7	18	3	13	10	20	11	8	14	5	16	9
18	1	7	8	15	14	6	19	11	20	9	4	17	2	16	3	10	13	12	5
14	19	9	7	17	18	4	11	12	8	6	10	3	16	13	1	20	2	5	15
9	11	18	1	8	4	16	10	7	13	19	12	5	17	6	20	2	3	14	15
14	10	8	9	4	20	6	11	17	1	16	3	18	13	7	15	19	12	2	5
17	15	18	16	6	14	8	9	7	11	2	10	19	3	5	20	4	1	13	12
5	7	2	20	12	13	6	19	3	1	15	8	17	18	10	11	16	4	14	9
12	4	10	8	14	16	11	1	19	13	18	15	9	17	6	3	2	7	20	5
5	14	8	11	9	4	12	20	2	1	15	6	10	17	13	3	7	18	19	16
6	3	12	19	14	15	13	2	17	1	5	16	18	4	11	20	9	8	7	10
18	11	6	10	16	1	8	12	15	3	14	20	4	2	9	13	5	19	17	7
18	12	4	7	11	13	5	9	2	14	19	3	1	17	8	16	6	15	20	10
16	5	20	11	2	9	15	1	19	3	10	14	12	6	7	17	8	4	13	18
6	20	11	13	18	14	17	8	12	7	5	4	2	16	9	10	15	3	19	1
8	6	20	16	11	12	15	3	9	7	13	1	5	14	4	18	2	10	17	19
18	12	5	16	1	9	7	6	8	4	13	14	19	2	15	10	3	11	17	20
4	15	9	20	2	10	5	13	6	3	7	16	1	8	18	14	17	19	12	11
16	19	17	13	15	12	11	18	8	12	5	10	20	7	3	1	14	6	9	4
12	4	2	18	5	11	1	7	17	6	14	8	9	3	16	10	15	13	20	19

Appendix Table 4—*continued*

13	20	9	5	1	6	8	4	7	19	16	15	10	2	12	17	11	18	14	3
6	7	11	8	10	1	20	16	12	9	13	15	14	18	3	17	2	5	4	19
17	8	11	10	4	5	12	9	7	1	19	14	20	6	2	18	15	3	16	13
5	16	9	18	6	15	19	3	4	20	14	1	2	12	7	17	10	13	8	11
8	11	15	16	1	18	12	10	5	13	20	2	14	17	9	6	4	7	3	19
17	6	10	12	2	4	9	18	5	13	8	15	11	1	16	3	20	7	14	19
20	17	5	2	9	7	10	11	8	1	4	19	14	16	12	3	13	15	6	18
9	12	8	20	15	5	10	2	16	17	11	18	6	3	4	1	14	7	13	19
19	15	14	11	5	18	16	9	17	2	20	7	12	4	13	6	1	8	3	10
1	10	8	17	6	19	18	7	12	11	13	5	14	2	20	4	9	16	3	15
4	3	13	19	6	1	5	12	17	8	11	2	16	20	10	18	9	15	14	7
11	7	12	10	16	5	18	4	1	13	19	2	3	14	6	8	9	20	17	15
7	14	13	3	15	19	16	8	5	20	11	6	12	17	1	10	18	9	2	4
14	6	4	10	2	8	19	5	20	15	9	3	17	12	16	1	18	13	7	11
13	12	17	7	4	5	2	19	9	20	15	11	6	3	1	14	16	18	8	10
17	12	10	15	18	4	14	6	16	3	13	11	1	5	19	8	20	9	2	7
16	20	11	2	8	18	4	14	7	12	3	5	15	19	13	10	6	17	9	1
14	6	8	17	4	13	7	5	9	20	12	10	11	19	1	15	16	2	18	3
20	18	2	1	3	12	14	7	10	11	6	17	15	16	5	9	8	13	4	19
19	3	1	17	4	5	13	10	7	9	11	18	2	12	8	20	15	6	14	16
7	20	1	15	19	16	12	3	5	11	4	13	6	10	8	9	17	2	14	18
9	7	12	4	5	19	18	15	2	11	1	17	8	16	20	13	3	6	14	10
12	10	5	3	19	1	2	14	6	16	9	8	15	11	13	4	20	17	18	7
8	15	13	4	9	11	17	18	7	6	12	16	14	5	2	3	1	20	10	19
10	6	1	3	16	12	14	11	17	4	8	9	5	19	2	7	18	15	13	20
16	14	8	12	1	10	19	5	2	17	11	15	7	3	4	13	18	20	9	6
9	12	19	18	16	5	14	20	15	8	13	11	10	7	3	2	6	1	17	4
5	10	8	12	9	18	17	6	13	2	1	3	15	20	7	16	19	11	4	14
12	9	4	1	3	8	10	18	13	14	2	6	15	7	20	19	17	16	5	11
7	18	2	10	15	5	9	17	3	20	16	6	12	14	8	1	11	19	4	13
17	2	5	12	8	10	14	16	7	3	20	11	4	1	9	19	18	15	6	13
8	20	13	10	9	19	15	2	7	17	14	16	6	11	4	12	1	5	3	18
13	15	4	7	1	8	14	12	2	6	20	9	16	5	18	19	3	10	11	17
16	6	13	4	2	19	8	7	1	9	14	11	15	18	12	5	20	10	3	17
5	15	7	16	13	20	2	14	8	19	10	6	9	3	11	12	17	18	1	4
10	14	8	5	6	16	2	7	11	9	18	3	1	13	15	12	17	19	20	4
5	2	16	11	6	3	17	10	19	9	14	8	20	13	15	1	12	7	18	4
9	16	4	10	7	17	11	15	8	18	19	3	14	1	5	20	6	2	12	13
4	9	11	10	5	19	20	6	7	3	13	8	15	17	1	14	16	18	2	12
16	14	15	2	18	5	12	3	17	1	6	9	7	8	20	10	11	13	4	19
4	16	6	11	17	18	1	14	2	15	20	12	7	19	8	9	13	10	5	3
19	14	2	17	3	12	1	18	6	10	13	7	5	4	8	11	9	15	16	20
10	7	1	9	19	18	15	12	3	20	4	6	17	5	8	14	11	16	13	2
11	7	16	3	15	9	19	20	12	6	4	10	13	2	18	1	8	14	17	5
18	9	10	11	14	12	6	19	5	4	2	13	15	20	8	1	16	3	7	17

Appendix Table 4—*continued*

2	17	12	9	18	20	11	7	8	3	19	1	16	15	4	6	5	14	13	10
9	4	2	10	14	5	7	17	8	18	20	6	16	15	19	13	1	12	11	3
6	13	18	9	20	2	17	3	10	12	19	5	7	4	8	1	15	14	11	16
14	4	16	13	10	5	7	17	20	12	1	6	2	11	15	19	3	18	9	8
9	15	19	3	5	8	14	11	1	4	18	20	16	7	10	17	12	2	6	13
7	20	10	2	19	15	5	6	18	1	9	11	14	4	3	13	16	12	17	8
17	4	10	19	12	8	5	16	6	18	15	2	11	7	13	9	20	3	14	1
16	14	19	15	9	18	8	13	3	11	2	20	4	12	10	6	17	7	1	5
6	18	15	13	11	8	10	12	1	19	14	17	2	9	16	3	20	4	7	5
16	15	13	1	19	12	10	18	7	3	6	20	2	4	9	11	14	17	8	5
1	9	20	8	4	2	7	5	16	17	18	6	14	11	19	12	15	10	13	3
13	4	1	9	6	8	3	2	18	19	17	20	10	15	14	11	7	5	12	16
6	15	3	1	16	19	13	2	17	10	11	5	20	7	4	14	8	9	12	18
16	1	10	11	15	5	14	18	12	3	19	7	8	9	4	6	17	20	2	13
1	14	10	7	2	8	15	5	20	11	17	9	19	18	6	16	4	3	12	13
2	19	11	16	1	5	18	9	3	20	12	17	15	10	4	6	7	8	13	14
18	15	13	17	2	4	11	1	14	6	16	20	7	12	3	8	10	19	9	5
10	14	9	7	12	8	1	2	19	20	13	4	17	16	6	18	3	11	15	5
5	13	6	3	19	20	16	12	4	18	9	10	15	17	1	8	7	14	11	2
6	13	20	11	2	16	18	12	3	5	8	10	19	9	15	14	1	7	4	17
10	6	1	2	5	12	8	3	7	9	11	19	16	18	4	14	13	20	17	15
7	2	20	18	10	3	5	4	9	14	16	17	6	13	12	19	15	1	8	11
15	11	17	19	2	9	16	7	4	20	18	6	1	12	5	8	10	3	13	14
20	5	4	9	15	11	2	17	13	6	12	1	14	8	16	10	19	7	3	18
5	13	19	12	15	20	4	2	17	9	7	10	14	6	16	8	3	11	18	1
12	13	5	6	8	9	7	15	2	18	20	4	14	3	11	16	17	19	1	10
6	2	10	18	9	16	20	13	17	5	4	12	1	11	7	19	15	8	3	14
15	11	10	8	20	6	7	1	9	4	2	14	18	19	3	16	13	12	5	17
17	15	19	13	7	12	20	8	14	16	9	11	6	2	3	1	4	18	5	10
7	2	9	19	3	18	6	13	10	20	11	12	4	17	16	8	1	5	14	15
2	16	4	6	17	18	14	9	12	3	1	5	15	8	19	10	11	20	13	7
17	3	2	20	12	18	19	13	6	10	14	16	1	4	11	7	9	8	5	15
8	5	2	17	10	15	12	18	1	20	14	3	13	11	19	4	6	9	16	7
11	19	20	1	6	13	2	7	3	17	4	8	12	16	15	10	18	9	5	14
17	7	6	1	8	15	18	20	5	10	12	4	3	9	16	19	2	13	14	11
12	6	9	14	2	17	4	20	16	10	7	15	1	8	19	13	5	18	3	11
11	17	9	19	2	5	15	20	6	16	4	1	10	7	8	3	13	18	12	14
20	11	5	1	9	17	10	19	7	2	4	15	16	14	6	12	13	3	18	8
11	17	20	1	10	2	12	18	3	6	4	13	16	19	7	9	14	15	8	5
6	3	10	15	11	4	20	2	5	18	9	19	12	7	1	16	17	8	14	13
2	4	12	16	3	15	6	5	11	10	17	19	18	9	7	18	14	1	20	8
14	8	16	4	13	19	6	11	3	9	1	7	2	10	17	15	5	20	18	12
17	1	8	12	15	5	4	16	20	19	6	2	7	14	9	11	3	18	10	13
5	6	13	9	4	19	16	17	15	11	2	10	7	8	12	3	1	20	14	18
13	19	20	2	7	9	15	18	4	6	8	1	5	14	16	3	11	17	12	10

Appendix Table 4—*continued*

2	4	14	10	13	18	11	19	15	16	7	9	3	8	6	17	12	5	20	1
9	1	13	18	16	8	15	17	4	6	7	20	14	5	10	12	3	11	2	19
5	2	11	7	1	16	20	14	3	4	15	13	9	8	12	17	19	6	10	18
8	16	18	4	10	20	2	19	5	7	13	1	11	17	15	12	6	9	14	3
9	6	16	12	18	19	17	8	10	5	3	4	13	20	1	15	2	11	14	7
2	13	4	9	14	8	10	3	16	17	5	20	1	7	19	11	6	18	15	12
20	15	8	11	12	18	5	19	3	4	9	13	17	1	2	10	6	16	7	14
13	16	19	20	15	5	3	4	11	2	17	12	14	7	8	1	10	9	6	18
15	7	20	16	1	14	5	3	12	17	19	4	9	2	11	8	10	18	6	13
7	15	16	5	6	14	11	4	20	10	12	1	18	19	9	17	8	3	2	13
14	18	13	1	8	3	11	7	20	15	16	5	19	9	12	10	4	2	17	6
8	12	7	9	6	14	10	5	16	13	4	19	20	17	11	1	18	2	15	3
14	6	9	3	2	18	8	16	13	12	10	4	15	5	17	1	20	7	19	11
6	3	15	13	2	1	4	18	10	7	19	11	12	16	5	17	8	20	9	14
7	14	3	8	13	2	15	6	16	19	20	1	18	5	4	12	17	11	10	9
11	12	6	5	9	10	3	20	4	16	1	18	15	19	7	8	17	2	13	14
4	20	17	7	10	9	1	3	16	8	6	2	18	14	19	12	15	5	11	13
2	18	20	6	4	3	7	16	17	13	12	5	14	15	19	11	9	10	1	8
8	3	13	18	11	9	4	20	1	14	7	16	12	17	6	15	10	5	19	2
20	18	11	1	15	7	12	2	10	6	17	14	8	19	9	13	16	5	3	4
5	7	2	8	6	10	11	17	12	16	18	15	4	19	20	9	13	1	14	3
13	20	16	17	9	19	4	15	7	10	14	3	2	11	12	18	6	8	5	1
14	3	13	16	5	2	15	9	17	19	12	6	18	7	11	20	10	8	1	4
13	9	11	6	3	2	14	16	18	15	10	20	4	1	7	8	12	17	5	19
4	7	1	11	5	18	15	6	10	13	9	14	3	19	2	16	20	17	12	8
6	18	2	9	19	3	17	1	13	4	12	20	15	14	7	11	8	16	5	10
20	7	12	13	2	1	6	11	14	5	19	9	4	15	18	8	10	16	3	17
5	12	4	20	13	3	8	17	19	16	18	14	2	10	15	9	1	11	7	6
1	8	2	17	7	11	12	20	14	5	10	16	15	3	9	4	19	18	13	6
10	2	18	11	13	12	8	4	14	16	17	3	20	5	15	9	6	1	7	19
5	16	9	3	13	14	2	15	17	8	4	6	10	18	11	1	19	12	20	7
7	19	12	18	15	2	8	16	4	17	11	20	3	5	14	9	10	6	13	1
18	12	2	19	1	16	3	5	6	15	17	11	10	20	8	4	7	14	9	13
3	14	11	18	7	8	6	12	1	2	5	19	4	15	20	9	17	13	10	16
16	11	8	18	4	7	6	14	12	17	5	9	19	10	15	13	2	1	3	20
17	11	15	13	1	14	12	6	9	18	4	19	7	16	3	20	5	2	10	8
1	20	8	15	14	10	13	9	12	19	17	18	11	6	4	3	5	2	7	16
17	7	15	2	18	12	5	11	9	1	16	6	4	20	13	3	19	14	10	8
4	16	1	11	15	9	13	3	6	7	10	20	12	19	8	18	2	14	17	5
1	16	8	13	10	19	12	15	11	17	18	5	2	4	20	9	3	7	6	14
4	16	12	9	11	7	5	2	13	14	1	19	3	15	20	18	6	10	8	17
1	14	15	18	19	13	3	12	11	16	10	9	7	4	8	20	5	17	2	6
7	1	15	14	6	11	3	2	12	4	10	19	9	5	18	17	20	16	13	8
7	20	8	18	17	2	16	9	12	1	4	13	3	15	11	14	19	6	10	5
10	1	16	6	18	2	4	15	3	9	11	17	12	5	7	19	8	13	14	20

Appendix Table 4—*continued*

6	8	16	17	11	13	4	1	20	2	12	9	7	18	5	10	19	15	14	3
8	5	7	2	20	1	11	19	12	16	10	13	17	4	3	15	18	6	14	9
20	14	4	7	18	13	12	11	17	16	19	5	3	9	10	8	2	1	6	15
8	16	11	15	3	20	9	6	13	14	17	12	7	5	2	4	19	1	18	10
4	3	17	18	12	5	9	2	8	14	7	6	16	15	10	19	11	20	13	1
5	16	10	11	19	4	12	1	3	9	2	20	6	14	18	8	13	17	15	7
10	5	15	17	9	14	7	16	11	12	19	3	8	20	4	13	6	1	2	18
11	15	8	14	9	5	13	18	7	17	4	6	3	10	16	12	20	1	2	19
1	13	19	17	2	11	16	6	9	18	4	3	8	12	5	10	7	15	14	20
9	10	2	13	18	8	4	5	3	16	15	7	14	19	17	20	11	12	6	1
20	4	12	13	3	5	14	17	10	15	2	9	16	1	18	19	11	7	6	8
18	8	14	17	9	1	11	10	12	4	13	6	3	16	7	19	20	5	2	15
5	20	9	13	17	14	19	2	8	7	1	6	12	4	18	15	10	3	11	16
18	8	2	12	1	7	16	19	5	9	13	3	20	4	6	11	15	10	17	14
12	4	1	15	10	11	18	19	7	13	17	9	16	8	14	20	2	5	3	6
6	20	3	10	13	19	14	12	4	7	2	1	9	18	11	5	15	16	8	17
7	9	16	15	17	1	11	13	14	5	6	2	19	18	4	3	8	12	20	10
13	5	19	3	10	8	20	9	1	2	6	18	7	14	15	17	4	12	11	16
15	2	7	6	4	17	8	1	13	11	10	18	20	5	12	19	3	9	14	16
18	7	16	6	2	12	3	13	9	15	8	11	5	1	19	20	10	14	4	17
19	5	15	7	10	2	16	11	8	4	14	1	9	12	20	17	18	13	6	3
3	11	1	2	10	12	5	19	8	18	16	14	13	9	15	17	4	6	20	7
18	7	14	5	13	2	1	8	20	9	19	17	4	3	11	16	6	15	12	10
11	15	16	17	7	4	9	19	5	3	8	18	10	20	14	6	2	12	13	1
6	17	20	9	14	12	1	10	15	19	5	11	16	13	3	7	4	2	18	8
1	16	6	3	15	13	17	14	12	20	11	10	4	9	19	18	7	8	5	2
2	1	13	17	14	15	3	20	11	18	6	12	19	16	7	4	5	10	8	9
20	8	14	2	18	11	16	1	13	12	6	4	15	7	3	19	9	10	5	17
3	9	4	8	12	5	19	6	15	17	10	2	20	16	13	18	14	11	7	1
12	9	18	1	10	4	16	8	7	20	17	2	11	6	13	14	3	15	5	19
18	13	11	10	3	17	19	20	8	15	16	1	5	7	6	12	9	4	14	2
18	9	4	14	11	17	2	1	8	5	15	16	20	7	6	10	13	12	19	3
9	11	5	4	12	7	1	19	16	18	20	13	10	2	3	6	14	15	8	17
14	15	6	20	7	8	19	13	1	5	4	16	9	3	10	11	18	17	12	2
19	15	10	1	12	16	9	8	13	2	3	6	20	11	7	5	4	17	14	18
20	5	17	7	14	1	6	16	4	2	13	9	3	10	8	12	15	19	11	18
7	9	12	8	20	11	6	16	2	3	4	15	1	17	19	5	14	13	18	10
11	15	9	6	18	3	2	19	13	20	12	16	14	4	17	5	7	10	8	1
4	3	17	14	2	18	7	12	16	10	11	15	5	20	13	1	8	9	19	6
8	6	17	4	10	5	9	3	18	20	15	16	13	1	11	12	14	7	19	2
11	5	15	9	1	12	6	16	17	3	20	13	14	18	8	7	19	10	4	2
4	12	13	9	1	7	15	3	6	19	2	8	16	11	17	18	5	20	10	14
8	15	5	4	13	7	19	10	20	12	17	2	11	6	9	16	3	18	1	14
3	12	11	18	2	15	13	14	8	19	7	20	10	16	9	6	1	17	4	5
19	1	10	6	9	2	20	5	18	17	11	16	8	14	4	12	15	13	7	3

Appendix Table 4—*continued*

```
 5  2  1  7 14 17  4 16  9  3 13 19 15 18 10 12 20  8 11  6
 2 12 18 13  4 10  5  6 14 16 15 19  9  7  3 17  1 11 20  8
15 13  2 18  6 19  7 10 11 16 20  1 12  3 14  9 17  4  8  5
 4 12 17  1  7  9 16  8 14 20 10 13  6  2 19  3 15 18 11  5
14 17 16  7 19  9  8  3 15 10  5 13 11 18  2 20  4  1 12  6

15  8 13 19  6 18  5  7 10 11  3 14  2  1 16  9  4 12 17 20
20  8 12 11  7 10  2 13  4 16 19 15  1  9 17  6  3 14 18  5
11  4  9  5 15 10 19 12 13  7  2  6  3 16 17 14 18  8  1 20
 9 19 15 13  2 12  8 14  7  4 20  6  1  5 11  3 17 18 16 10
 1  2 17 13  3  7  9  5  6 20 14 10  4 12 18 19  8 11 15 16

18 12 16 19 17  9  5 11  6 13  8 14  3  2 15 20  7  1 10  4
11 16 19  6 12  7  2 14 15 18  4  8  5  3  9 17 20  1 13 10
 4 19 12 17  7  6 14 20  5 11  9 13 18  8 10  2  3 15  1 16
 3 14  7 13 17  6  9  8 15 19  2 20 12 11 18  4 10  1  5 16
 6  4  3 13 12 10 11  2 15  7  9 20  8  1 19 18  5 17 16 14

 3  6  4 13 19  1  7 15  8 10 12 18 11  9 20  5  2 16 14 17
10  6 20 14 19  4  1  8 16  2 13  3  7  9 12  5 11 15 17 18
17  7  8 16  4  6  9 19 11  5 12 13 10 18 20  1 14 15  2  3
12 16  9  3  4 18 19 10  8  6 20 14 13  7  5 15 11  2 17  1
 3  5 15  9  6 19  4 17  2 14 11 20 13 10  7 18 12  8 16  1

 8  7  2 12 17 19  9  5 20 13 11 18  1  3  4 15 10 14 16  6
20 13  4  6  7  9  2 16 11 14  5 19  3 17 12 18 10  1  8 15
14 19  5  6  3 16 10  8 12 17 20  9 13 18  2 11  7  4 15  1
19  3 12 11 13 14  7 18  5  1  2 17 15  6  9 20  4  8 16 10
 5 11 15 17  9  6  1 10 20 19 16  8 14  2  7  3 13 18 12  4

14  3 11  2  6  1 15 19  8  9 12 10 16  7  5 17 20 13  4 18
 9 15 18  6 14 13 12  2  1  4  8  3 19  5 20 11  7 10 16 17
20 13 16 18  3  9 14  5  8 12 17 10 15  7 19  2 11  6  4  1
 8 19 15 20  9  1 18  7 17 14  4 16 11  5  6 10 13  2 12  3
18 15 16  1 20 14 11 13  9  2  5  8 10 17 12  7 19  4  3  6

16  2 15  3  8 11 12  1 14 19  7 17 20  4  6  9 10 13  5 18
 6  9  4  1  2 15 18 16  8 17 13  3 14 20 12  5 19  7 10 11
12 10  4  5 16  1  7 19  8 11  2  3 14  6 18 13 15  9 20 17
17 12 14  7  9  6 19  8 16  5 11  1 20 15  3  4 18 13 10  2
 5  4 10 13 15 20  7  3 12  9 14  1 17 19  6 16 18  2  8 11

10  8  2  9 16  7 11 20 17 19  5 18 13  6  4 15  1 12  3 14
10  3  9 14  5  8  1  6 17  7 15 12  2  4 16 18 13 20 11 19
14  8 16  6 19  7  5 18  4 17  2 20  1 13 12 15  9 11  3 10
20 14  3 11  5  8 19 18  4  7 17  2 16 13 12 10  1 15  9  6
 1  9 19 16 11 15  7  8 14 10 12 18  4  2  6 20 17 13  5  3

 6  3  7 18 17  1  5  4  8 14 20  2 12 15 10 19 11 13  9 16
18 13  4 16  1  3  6  9 11  8 19  5 17  2 20  7 14 15 12 10
13 12  3 18 19 10  6  9 14 15 20  7 17  2  8 11 16  5  4  1
12 18  5  1 16 11  6  4  2 17 14  3 20 10  7 13 19  8 15  9
12  6 13  4  2  1 18 20 14  9  3 16  8 17 19 10  5 11 15  7
```

Appendix Table 4—*continued*

18	3	4	7	15	14	16	1	10	12	6	13	20	8	2	19	9	17	11	5
10	9	8	4	5	13	19	12	2	6	20	1	15	18	16	7	14	3	11	17
15	16	17	4	11	20	10	2	12	3	13	19	6	18	7	8	1	5	9	14
5	18	11	16	10	3	13	7	4	12	8	6	17	2	15	20	19	14	1	9
13	10	17	6	12	9	3	5	16	18	19	7	1	8	2	20	14	11	15	4
5	6	19	8	3	17	20	4	2	18	15	14	10	7	11	16	9	1	13	12
2	7	10	17	15	3	18	5	11	12	4	14	9	20	13	16	1	6	19	8
8	2	10	13	1	11	15	6	9	19	17	16	20	7	12	5	18	14	4	3
8	1	3	15	10	2	7	13	19	9	12	11	18	17	20	16	14	4	6	5
15	2	19	1	16	18	13	8	5	14	4	7	11	3	17	12	9	10	20	6
20	14	18	5	2	10	11	7	15	16	13	17	12	9	8	6	4	3	19	1
18	13	6	9	1	5	19	2	16	17	10	15	7	11	20	14	8	4	3	12
9	13	20	16	11	8	12	15	7	14	4	10	6	17	2	3	18	5	19	1
7	10	4	8	3	1	14	15	2	6	17	20	11	5	19	12	18	13	9	16
16	18	12	7	13	14	4	10	15	11	6	9	1	5	17	19	3	20	8	2
20	8	11	2	5	19	15	3	6	12	10	18	4	16	14	9	13	1	17	7
12	15	16	8	7	2	18	4	6	1	19	20	3	10	5	17	13	14	11	9
11	13	2	10	3	17	1	7	15	5	8	6	12	18	4	9	14	20	19	16
1	8	5	12	9	16	11	20	3	10	7	18	2	17	14	19	6	4	15	13
14	10	17	19	5	11	7	6	1	12	16	2	20	15	3	8	13	9	4	18
20	6	10	8	12	3	11	9	7	13	14	16	2	15	19	18	17	4	5	1
3	6	1	19	18	5	7	15	13	9	17	8	16	2	10	4	20	14	12	11
9	1	12	20	17	4	11	10	16	6	18	5	2	15	8	7	19	14	13	3
17	11	6	16	9	7	20	5	15	8	12	1	18	4	19	13	2	3	14	10
7	11	6	13	19	12	16	10	17	5	8	20	3	18	2	4	14	9	15	1
18	3	2	16	20	1	14	15	12	6	19	8	5	9	10	7	11	4	13	17
17	9	16	10	13	12	5	4	7	20	19	11	15	6	1	3	2	14	18	8
1	7	17	3	15	6	20	2	8	10	12	5	18	4	11	19	14	16	9	13
15	6	18	17	20	8	3	14	5	4	7	13	10	2	1	19	9	12	16	11
8	7	1	17	10	5	13	16	3	18	15	12	4	2	11	9	19	14	6	20
16	20	15	1	13	4	17	12	19	2	5	8	18	11	9	7	3	6	14	10
2	13	12	3	5	4	9	10	17	14	8	16	11	18	6	15	20	19	7	1
7	11	20	5	10	18	8	12	6	4	19	13	1	15	17	9	3	16	14	2
6	13	17	10	2	15	5	19	20	7	3	11	14	4	9	18	12	1	8	16
6	18	9	17	14	15	1	8	13	11	7	20	2	4	10	19	5	16	12	3
4	2	18	9	7	1	5	17	16	3	20	13	12	8	19	10	6	15	11	14
18	1	17	3	15	6	4	14	20	12	9	2	19	8	7	13	16	11	5	10
5	8	1	18	16	17	13	12	2	7	10	15	6	11	4	19	14	20	9	3
5	18	11	1	9	19	3	10	4	15	8	2	20	6	14	12	7	17	13	16
13	1	6	9	4	15	12	10	11	3	20	18	5	8	17	19	7	16	14	2
15	14	11	19	6	10	12	8	4	18	13	20	9	3	17	16	2	7	5	1
13	12	16	5	10	4	19	15	18	1	2	8	20	17	7	3	9	11	6	14
8	2	16	10	18	4	9	6	12	5	15	17	20	3	1	14	11	13	7	19
8	13	6	12	9	14	20	7	4	19	10	5	11	17	1	18	3	15	2	16
3	8	11	2	18	15	9	1	10	13	20	5	14	4	17	19	6	7	12	16

Appendix Table 4—*concluded*

16	17	12	2	15	4	19	10	7	1	3	9	11	5	13	6	20	8	14	18
19	3	18	10	15	8	12	13	11	17	6	14	9	20	1	5	2	16	7	4
19	9	3	5	8	6	11	16	4	15	2	18	10	20	14	7	12	17	13	1
6	7	8	14	16	12	11	2	17	3	5	1	4	10	18	13	15	19	9	20
9	20	17	5	19	13	6	12	15	14	3	11	16	8	18	7	10	4	2	1
12	14	19	16	10	1	13	5	8	4	18	17	9	6	11	15	7	3	2	20
18	1	2	8	4	17	7	15	19	5	12	6	20	16	13	11	14	10	3	9
8	6	5	11	13	20	7	2	4	17	3	10	1	15	14	9	12	16	19	18
8	11	16	20	10	19	6	1	15	2	14	7	13	4	5	18	3	9	17	12
2	7	9	6	15	10	3	18	17	1	5	4	19	11	8	14	16	12	13	20
5	8	1	15	18	7	3	4	12	16	10	19	14	20	6	17	9	2	13	11
15	5	20	18	9	6	17	8	10	11	3	2	12	4	7	13	14	19	16	1
14	3	4	11	1	20	17	15	10	13	2	6	7	9	19	8	5	12	18	16
8	10	14	5	6	11	7	15	20	3	12	19	13	1	16	4	17	18	9	2
12	2	19	10	20	17	15	16	3	11	5	4	8	18	9	6	1	14	13	7
18	10	14	3	13	19	20	15	1	12	16	7	5	17	8	6	11	9	4	2
2	6	11	8	3	5	12	10	20	13	19	15	17	9	7	16	14	4	1	18
18	11	10	13	8	20	7	6	9	19	14	5	12	1	17	3	16	15	2	4
20	5	19	7	10	17	9	3	14	1	16	8	15	18	11	4	12	2	6	13
7	17	10	15	4	2	13	1	5	14	9	16	18	3	12	11	8	20	19	6
19	8	11	14	9	17	5	2	13	10	16	6	1	4	12	20	18	7	15	3
2	4	15	3	13	7	16	20	6	10	8	19	17	5	14	11	12	1	9	18
18	5	9	13	7	1	12	6	14	15	2	16	19	17	4	3	10	8	20	11
11	12	14	10	20	9	8	4	15	17	19	1	3	18	7	5	16	6	13	2
9	18	17	8	4	5	6	2	14	13	7	15	10	3	12	19	16	20	11	1
1	9	2	18	6	8	12	19	20	15	4	3	13	16	5	14	11	17	7	10
13	11	14	19	18	15	20	5	10	17	7	8	9	2	16	1	6	3	12	4
3	14	11	6	18	10	2	15	7	9	19	4	17	20	12	1	5	16	8	13
18	1	20	7	3	11	13	15	17	10	16	4	6	5	2	12	8	9	19	14
10	20	7	18	9	5	6	15	2	17	16	19	13	14	12	1	8	11	3	4
4	14	20	19	6	13	17	8	10	7	5	12	11	2	18	3	9	15	16	1
6	5	2	8	20	9	15	1	7	10	17	14	4	19	16	13	12	11	3	18
13	9	2	12	14	18	6	11	20	4	8	19	15	5	7	3	17	1	16	10
12	17	6	3	18	1	15	9	11	14	20	5	7	19	2	13	8	10	4	16
7	12	3	10	16	20	13	6	18	2	17	14	4	5	1	11	9	8	19	15
1	4	18	17	16	10	5	14	20	15	9	3	13	2	7	6	8	11	12	19
18	9	4	2	16	11	5	19	14	1	12	15	8	17	13	7	20	10	3	6
3	1	16	18	14	11	4	15	5	13	19	17	6	12	10	7	8	2	20	9
10	19	7	5	20	3	2	13	14	9	8	12	6	15	4	17	16	11	18	1
20	15	19	6	7	8	9	2	3	12	18	11	14	16	13	10	5	1	4	17
6	7	3	17	5	9	11	4	1	13	18	2	8	20	10	16	12	14	15	19
10	3	16	4	6	2	11	9	19	17	8	1	14	7	5	20	13	12	15	18
13	5	6	16	10	7	8	12	2	20	3	4	17	11	9	1	15	19	18	14
10	5	15	3	2	19	13	18	9	14	16	4	20	17	12	11	8	1	6	7
10	18	12	19	11	14	4	3	8	2	7	20	5	17	16	6	15	13	1	9

Appendix Tables 5

Probability function of S
(for Kendall's coefficient of concordance)

Each table entry is the probability that S (for Kendall's coefficient of concordance W) is greater than or equal to a specified positive value.

Table 5A. $n = 3$ and $m = 2$ to $m = 10$.

S	2	3	4	5	Values of m 6	7	8	9	10
0	1.000	1.000	1.000	1.000	1.000	1.000	1.000	1.000	1.000
2	0.833	0.944	0.931	0.954	0.956	0.964	0.967	0.971	0.974
6	0.500	0.528	0.653	0.691	0.740	0.768	0.794	0.814	0.830
8	0.167	0.361	0.431	0.522	0.570	0.620	0.654	0.685	0.710
14		0.194	0.273	0.367	0.430	0.486	0.531	0.569	0.601
18		0.028	0.125	0.182	0.252	0.305	0.355	0.398	0.436
24			0.069	0.124	0.184	0.237	0.285	0.328	0.368
26			0.042	0.093	0.142	0.192	0.236	0.278	0.316
32			0.0046	0.039	0.072	0.112	0.149	0.187	0.222
38				0.024	0.052	0.085	0.120	0.154	0.187
42				0.0085	0.029	0.051	0.079	0.107	0.135
50				$0.0^3 77$	0.012	0.027	0.047	0.069	0.092
54					0.0081	0.021	0.038	0.057	0.078
56					0.0055	0.016	0.030	0.048	0.066
62					0.0017	0.0084	0.018	0.031	0.046
72					$0.0^3 13$	0.0036	0.0099	0.019	0.030
74						0.0027	0.0080	0.016	0.026
78						0.0012	0.0048	0.010	0.018
86						$0.0^3 32$	0.0024	0.0060	0.012
96						$0.0^3 32$	0.0011	0.0035	0.0075
98						$0.0^4 21$	$0.0^3 86$	0.0029	0.0063
104							$0.0^3 26$	0.0013	0.0034
114							$0.0^4 61$	$0.0^3 66$	0.0020
122							$0.0^4 61$	$0.0^3 35$	0.0013
126							$0.0^4 61$	$0.0^3 20$	$0.0^3 83$
128							$0.0^5 36$	$0.0^4 97$	$0.0^3 51$
134								$0.0^4 54$	$0.0^3 37$
146								$0.0^4 11$	$0.0^3 18$
150								$0.0^4 11$	$0.0^3 11$
152								$0.0^4 11$	$0.0^4 85$
158								$0.0^4 11$	$0.0^4 44$
162								$0.0^6 60$	$0.0^4 20$
168									$0.0^4 11$
182									$0.0^5 21$
200									$0.0^7 99$

Table 5B. $n = 4$ and $m = 3, 5$.

S	$m = 3$	$m = 5$	S	$m = 5$
1	1.000	1.000	61	0.055
3	0.958	0.975	65	0.044
5	0.910	0.944	67	0.034
9	0.727	0.857	69	0.031
11	0.608	0.771	73	0.023
13	0.524	0.709	75	0.020
17	0.446	0.652	77	0.017
19	0.342	0.561	81	0.012
21	0.300	0.521	83	0.0087
25	0.207	0.445	85	0.0067
27	0.175	0.408	89	0.0055
29	0.148	0.372	91	0.0031
33	0.075	0.298	93	0.0023
35	0.054	0.260	97	0.0018
37	0.033	0.226	99	0.0016
41	0.017	0.210	101	0.0014
43	0.0017	0.162	105	$0.0^3 64$
45	0.0017	0.141	107	$0.0^3 33$
49		0.123	109	$0.0^3 21$
51		0.107	113	$0.0^3 14$
53		0.093	117	$0.0^4 48$
57		0.075	125	$0.0^5 30$
59		0.067		

Table 5C. $n = 4$ and $m = 2$, 4 and 6.

S	m = 2	m = 4	m = 6	S	m = 6
0	1.000	1.000	1.000	82	0.035
2	0.958	0.992	0.996	84	0.032
4	0.833	0.928	0.957	86	0.029
6	0.792	0.900	0.940	88	0.023
8	0.625	0.800	0.874	90	0.022
10	0.542	0.754	0.844	94	0.017
12	0.458	0.677	0.789	96	0.014
14	0.375	0.649	0.772	98	0.013
16	0.208	0.524	0.679	100	0.010
18	0.167	0.508	0.668	102	0.0096
20	0.042	0.432	0.609	104	0.0085
22		0.389	0.574	106	0.0073
24		0.355	0.541	108	0.0061
26		0.324	0.512	110	0.0057
30		0.242	0.431	114	0.0040
32		0.200	0.386	116	0.0033
34		0.190	0.375	118	0.0028
36		0.158	0.338	120	0.0023
38		0.141	0.317	122	0.0020
40		0.105	0.270	126	$0.001\mathsf{5}$
42		0.094	0.256	128	$0.0^3 90$
44		0.077	0.230	130	$0.0^3 87$
46		0.068	0.218	132	$0.0^3 73$
48		0.054	0.197	134	$0.0^3 65$
50		0.052	0.194	136	$0.0^3 40$
52		0.036	0.163	138	$0.0^3 36$
54		0.033	0.155	140	$0.0^3 28$
56		0.019	0.127	144	$0.0^3 24$
58		0.014	0.114	146	$0.0^3 22$
62		0.012	0.108	148	$0.0^3 12$
64		0.0069	0.089	150	$0.0^4 95$
66		0.0062	0.088	152	$0.0^4 62$
68		0.0027	0.073	154	$0.0^4 46$
70		0.0027	0.066	158	$0.0^4 24$
72		0.0016	0.060	160	$0.0^4 16$
74		$0.0^3 94$	0.056	162	$0.0^4 12$
76		$0.0^3 94$	0.043	164	$0.0^5 80$
78		$0.0^3 94$	0.041	170	$0.0^5 24$
80		$0.0^4 72$	0.037	180	$0.0^6 13$

Table 5D. $n = 5$ and $m = 3$.

S	m = 3	S	m = 3
0	1.000	44	0.236
2	1.000	46	0.213
4	0.988	48	0.172
6	0.972	50	0.163
8	0.941	52	0.127
10	0.914	54	0.117
12	0.845	56	0.096
14	0.831	58	0.080
16	0.768	60	0.063
18	0.720	62	0.056
20	0.682	64	0.045
22	0.649	66	0.038
24	0.595	68	0.028
26	0.559	70	0.026
28	0.493	72	0.017
30	0.475	74	0.015
32	0.432	76	0.0078
34	0.406	78	0.0053
36	0.347	80	0.0040
38	0.326	82	0.0028
40	0.291	86	0.0^390
42	0.253	90	0.0^469

Appendix Table 6

Significance points of S
(for Kendall's coefficient of concordance)

From Friedman (1940) by permission of the author and the editor of the *Annals of Mathematical Statistics*.

m	3	4	5	6	7	m	S
			n			Additional values for *n* = 3	
Values at 0.05 level of significance							
3			64.4	103.9	157.3	9	54.0
4		49.5	88.4	143.3	217.0	12	71.9
5		62.6	112.3	182.4	276.2	14	83.8
6		75.7	136.1	221.4	335.2	16	95.8
8	48.1	101.7	183.7	299.0	453.1	18	107.7
10	60.0	127.8	231.2	376.7	571.0		
15	89.8	192.9	349.8	570.5	864.9		
20	119.7	258.0	468.5	764.4	1158.7		
Values at 0.01 level of significance							
3			75.6	122.8	185.6	9	75.9
4		61.4	109.3	176.2	265.0	12	103.5
5		80.5	142.8	229.4	343.8	14	121.9
6		99.5	176.1	282.4	422.6	16	140.2
8	66.8	137.4	242.7	388.3	579.9	18	158.6
10	85.1	175.3	309.1	494.0	737.0		
15	131.0	269.8	475.2	758.2	1129.5		
20	177.0	364.2	641.2	1022.2	1521.9		

Significance points of Fisher's z-distribution

Reprinted from Table VI of Fisher (1934), by permission of the author and publishers.

Table 7A. Five per cent points of the distribution of z.

Values of v_2	Values of v_1									
	1	2	3	4	5	6	8	12	24	∞
1	2.5421	2.6479	2.6870	2.7071	2.7194	2.7276	2.7380	2.7484	2.7588	2.7693
2	1.4592	1.4722	1.4765	1.4787	1.4800	1.4808	1.4819	1.4830	1.4840	1.4851
3	1.1577	1.1284	1.1137	1.1051	1.0994	1.0953	1.0899	1.0842	1.0781	1.0716
4	1.0212	0.9690	0.9429	0.9272	0.9168	0.9093	0.8993	0.8885	0.8767	0.8639
5	0.9441	0.8777	0.8441	0.8236	0.8097	0.7997	0.7862	0.7714	0.7550	0.7368
6	0.8948	0.8188	0.7798	0.7558	0.7394	0.7274	0.7112	0.6931	0.6729	0.6499
7	0.8606	0.7777	0.7347	0.7080	0.6896	0.6761	0.6576	0.6369	0.6134	0.5862
8	0.8355	0.7475	0.7014	0.6725	0.6525	0.6378	0.6175	0.5945	0.5682	0.5371
9	0.8163	0.7242	0.6757	0.6450	0.6238	0.6080	0.5862	0.5613	0.5324	0.4979
10	0.8012	0.7058	0.6553	0.6232	0.6009	0.5843	0.5611	0.5346	0.5035	0.4657
11	0.7889	0.6909	0.6387	0.6055	0.5822	0.5648	0.5406	0.5126	0.4795	0.4387
12	0.7788	0.6786	0.6250	0.5907	0.5666	0.5487	0.5234	0.4947	0.4592	0.4156
13	0.7703	0.6682	0.6134	0.5783	0.5535	0.5350	0.5089	0.4785	0.4419	0.3957
14	0.7630	0.6594	0.6036	0.5677	0.5423	0.5233	0.4964	0.4649	0.4269	0.3782
15	0.7568	0.6518	0.5950	0.5585	0.5326	0.5131	0.4855	0.4532	0.4138	0.3628
16	0.7514	0.6451	0.5876	0.5505	0.5241	0.5042	0.4760	0.4428	0.4022	0.3490
17	0.7466	0.6393	0.5811	0.5434	0.5166	0.4964	0.4676	0.4337	0.3919	0.3366
18	0.7424	0.6341	0.5753	0.5371	0.5099	0.4894	0.4602	0.4255	0.3827	0.3253
19	0.7386	0.6295	0.5701	0.5315	0.5040	0.4832	0.4535	0.4182	0.3743	0.3151
20	0.7352	0.6254	0.5654	0.5265	0.4986	0.4776	0.4474	0.4116	0.3668	0.3057
21	0.7322	0.6216	0.5612	0.5219	0.4938	0.4725	0.4420	0.4055	0.3599	0.2971
22	0.7294	0.6182	0.5574	0.5178	0.4894	0.4679	0.4370	0.4001	0.3536	0.2892
23	0.7269	0.6151	0.5540	0.5140	0.4854	0.4636	0.4325	0.3950	0.3478	0.2818
24	0.7246	0.6123	0.5508	0.5106	0.4817	0.4598	0.4283	0.3904	0.3425	0.2749
25	0.7225	0.6097	0.5478	0.5074	0.4783	0.4562	0.4244	0.3862	0.3376	0.2685
26	0.7205	0.6073	0.5451	0.5045	0.4752	0.4529	0.4209	0.3823	0.3330	0.2625
27	0.7187	0.6051	0.5427	0.5017	0.4723	0.4499	0.4176	0.3786	0.3287	0.2569
28	0.7171	0.6030	0.5403	0.4992	0.4696	0.4471	0.4146	0.3752	0.3248	0.2516
29	0.7155	0.6011	0.5382	0.4969	0.4671	0.4444	0.4117	0.3720	0.3211	0.2466
30	0.7141	0.5994	0.5362	0.4947	0.4648	0.4420	0.4090	0.3691	0.3176	0.2419
60	0.6933	0.5738	0.5073	0.4632	0.4311	0.4064	0.3702	0.3255	0.2654	0.1644
∞	0.6729	0.5486	0.4787	0.4319	0.3974	0.3706	0.3309	0.2804	0.2085	0

Table 7B.   One per cent points of the distribution of *z*.

Values of v_2	Values of v_1									
	1	2	3	4	5	6	8	12	24	∞
1	4.1535	4.2585	4.2974	4.3175	4.3297	4.3379	4.3482	4.3585	4.3689	4.3794
2	2.2950	2.2976	2.2984	2.2988	2.2991	2.2992	2.2994	2.2997	2.2999	2.3001
3	1.7649	1.7140	1.6915	1.6786	1.6703	1.6645	1.6569	1.6489	1.6404	1.6314
4	1.5270	1.4452	1.4075	1.3856	1.3711	1.3609	1.3473	1.3327	1.3170	1.3000
5	1.3943	1.2929	1.2449	1.2164	1.1974	1.1838	1.1656	1.1457	1.1239	1.0997
6	1.3103	1.1955	1.1401	1.1068	1.0843	1.0680	1.0460	1.0218	0.9948	0.9643
7	1.2526	1.1281	1.0672	1.0300	1.0048	0.9864	0.9614	0.9335	0.9020	0.8658
8	1.2106	1.0787	1.0135	0.9734	0.9459	0.9259	0.8983	0.8673	0.8319	0.7904
9	1.1786	1.0411	0.9724	0.9299	0.9006	0.8791	0.8494	0.8157	0.7769	0.7305
10	1.1535	1.0114	0.9399	0.8954	0.8646	0.8419	0.8104	0.7744	0.7324	0.6816
11	1.1333	0.9874	0.9136	0.8674	0.8354	0.8116	0.7785	0.7405	0.6958	0.6408
12	1.1166	0.9677	0.8919	0.8443	0.8111	0.7864	0.7520	0.7122	0.6649	0.6061
13	1.1027	0.9511	0.8737	0.8248	0.7907	0.7652	0.7295	0.6882	0.6386	0.5761
14	1.0909	0.9370	0.8581	0.8082	0.7732	0.7471	0.7103	0.6675	0.6159	0.5500
15	1.0807	0.9249	0.8448	0.7939	0.7582	0.7314	0.6937	0.6496	0.5961	0.5269
16	1.0719	0.9144	0.8331	0.7814	0.7450	0.7177	0.6791	0.6339	0.5786	0.5064
17	1.0641	0.9051	0.8229	0.7705	0.7335	0.7057	0.6663	0.6199	0.5630	0.4879
18	1.0572	0.8970	0.8138	0.7607	0.7232	0.6950	0.6549	0.6075	0.5491	0.4712
19	1.0511	0.8897	0.8057	0.7521	0.7140	0.6854	0.6447	0.5964	0.5366	0.4560
20	1.0457	0.8831	0.7985	0.7443	0.7058	0.6768	0.6355	0.5864	0.5253	0.4421
21	1.0408	0.8772	0.7920	0.7372	0.6984	0.6690	0.6272	0.5773	0.5150	0.4294
22	1.0363	0.8719	0.7860	0.7309	0.6916	0.6620	0.6196	0.5691	0.5056	0.4176
23	1.0322	0.8670	0.7806	0.7251	0.6855	0.6555	0.6127	0.5615	0.4969	0.4068
24	1.0285	0.8626	0.7757	0.7197	0.6799	0.6496	0.6064	0.5545	0.4890	0.3967
25	1.0251	0.8585	0.7712	0.7148	0.6747	0.6442	0.6006	0.5481	0.4816	0.3872
26	1.0220	0.8548	0.7670	0.7103	0.6699	0.6392	0.5952	0.5422	0.4748	0.3784
27	1.0191	0.8513	0.7631	0.7062	0.6655	0.6346	0.5902	0.5367	0.4685	0.3701
28	1.0164	0.8481	0.7595	0.7023	0.6614	0.6303	0.5856	0.5316	0.4626	0.3624
29	1.0139	0.8451	0.7562	0.6987	0.6576	0.6263	0.5813	0.5269	0.4570	0.3550
30	1.0116	0.8423	0.7531	0.6954	0.6540	0.6226	0.5773	0.5224	0.4519	0.3481
60	0.9784	0.8025	0.7086	0.6472	0.6028	0.5687	0.5189	0.4574	0.3746	0.2352
∞	0.9462	0.7636	0.6651	0.5999	0.5522	0.5152	0.4604	0.3908	0.2913	0

Appendix Table 8

Significance points of χ^2

Reproduced from Table III of Fisher (1934) by permission of the author and publishers.

ν	$P=0.99$	0.98	0.95	0.90	0.80	0.70	0.50	0.30	0.20	0.10	0.05	0.02	0.01
1	0.0³157	0.0³628	0.0²393	0.0158	0.0642	0.148	0.455	1.074	1.642	2.706	3.841	5.412	6.635
2	0.0201	0.0404	0.103	0.211	0.446	0.713	1.386	2.408	3.219	4.605	5.991	7.824	9.210
3	0.115	0.185	0.352	0.584	1.005	1.424	2.366	3.665	4.642	6.251	7.815	9.837	11.345
4	0.297	0.429	0.711	1.064	1.649	2.195	3.357	4.878	5.989	7.779	9.488	11.668	13.277
5	0.554	0.752	1.145	1.160	2.343	3.000	4.351	6.064	7.289	9.236	11.070	13.388	15.086
6	0.872	1.134	1.635	2.204	3.070	3.828	5.348	7.231	8.558	10.645	12.592	15.033	16.812
7	1.239	1.564	2.167	2.833	3.822	4.671	6.346	8.383	9.803	12.017	14.067	16.622	18.475
8	1.646	2.032	2.733	3.490	4.594	5.527	7.344	9.524	11.030	13.362	15.507	18.168	20.090
9	2.088	2.532	3.325	4.168	5.380	6.393	8.343	10.656	12.242	14.684	16.919	19.679	21.666
10	2.358	3.059	3.940	4.865	6.179	7.267	9.342	11.781	13.442	15.987	18.307	21.161	23.209
11	3.053	3.609	4.575	5.578	6.989	8.148	10.341	12.899	14.631	17.275	19.675	22.618	24.725
12	3.571	4.178	5.226	6.304	7.807	9.034	11.340	14.011	15.821	18.549	21.026	24.054	26.217
13	4.107	4.765	5.892	7.042	8.634	9.926	12.340	15.119	16.985	19.812	22.362	25.472	27.688
14	4.660	5.368	6.571	7.790	9.467	10.821	13.339	16.222	18.151	21.064	23.685	26.873	29.141
15	5.229	5.985	7.261	8.547	10.307	11.721	14.339	17.322	19.311	22.307	24.996	28.259	30.578
16	5.812	6.614	7.962	9.312	11.152	12.624	15.338	18.418	20.465	23.542	26.296	29.633	32.000
17	6.408	7.255	8.672	10.085	12.002	13.531	16.338	19.511	21.615	24.769	27.587	30.995	33.409
18	7.015	7.906	9.390	10.865	12.857	14.440	17.338	20.601	22.760	25.989	28.869	32.346	34.805
19	7.633	8.567	10.117	11.651	13.716	15.352	18.338	21.689	23.900	27.204	30.144	33.687	36.191
20	8.260	9.237	10.851	12.443	14.578	16.266	19.337	22.775	25.038	28.412	31.410	35.020	37.566
21	8.897	9.915	11.591	13.240	15.445	17.182	20.337	23.858	26.171	29.615	32.671	36.343	38.932
22	9.542	10.600	12.338	14.041	16.314	18.101	21.337	24.939	27.301	30.813	33.924	37.659	40.289
23	10.196	11.293	13.091	14.848	17.187	19.021	22.337	26.018	28.429	32.007	35.172	38.968	41.638
24	10.856	11.992	13.848	15.659	18.062	19.943	23.337	27.096	29.553	33.196	36.415	40.270	42.980
25	11.524	12.697	14.611	16.473	18.940	20.867	24.337	28.172	30.675	34.382	37.652	41.566	44.314
26	12.198	13.409	15.379	17.292	19.820	21.792	25.336	29.246	31.795	35.563	38.885	42.856	45.642
27	12.879	14.125	16.151	18.114	20.703	22.719	26.336	30.319	32.912	36.741	40.113	44.140	46.963
28	13.565	14.847	16.928	18.939	21.588	23.647	27.336	31.391	34.027	37.916	41.337	45.419	48.278
29	14.256	15.574	17.708	19.768	22.475	24.577	28.336	32.461	35.139	39.087	42.557	46.693	49.588
30	14.953	16.306	18.493	20.599	23.364	25.508	29.336	33.530	36.250	40.256	43.773	47.962	50.892

Note: For values of $\nu > 30$, $\sqrt{2\chi^2}$ is approximately normally distributed with mean $\sqrt{(2\nu - 1)}$ and variance one.

Appendix Table 9

Probability function of *d* in paired comparisons

This table gives the frequencies *f* for the number *d* of circular triads in paired comparisons of *n* objects, and the corresponding probability *P* that these values will be attained or exceeded. The entries labelled P' for $n = 10$ are approximate probabilities based on the chi-square approximation.

Value of d	n = 2 f	P	n = 3 f	P	n = 4 f	P	n = 5 f	P	n = 6 f	P	n = 7 f	P
0	2	1.000	6	1.000	24	1.000	120	1.000	720	1.000	5 040	1.000
1			2	0.250	16	0.625	120	0.883	960	0.978	8 400	0.998
2					24	0.375	240	0.766	2 240	0.949	21 840	0.994
3							240	0.531	2 880	0.880	33 600	0.983
4							280	0.297	6 240	0.792	75 600	0.967
5							24	0.023	3 648	0.602	90 384	0.931
6									8 640	0.491	179 760	0.888
7									4 800	0.227	188 160	0.802
8									2 640	0.081	277 200	0.713
9											280 560	0.580
10											384 048	0.447
11											244 160	0.263
12											233 520	0.147
13											72 240	0.036
14											2 640	0.001
Total	2	—	8	—	64	—	1 024	—	32 768	—	2 097 152	

Appendix Table 9—*continued*

Value of d	n = 8 f	P	n = 9 f	P	n = 10 f	P	P'
0	40 320	1.000	362 880	1.000	3 628 800	1.0^5	1.000
1	80 640	0.9^385	846 720	0.9^5	9 676 800	1.0^5	1.000
2	228 480	0.9^355	2 580 480	0.9^48	31 449 600	1.0^5	1.000
3	403 200	0.9^287	5 093 760	0.9^44	68 275 200	1.0^5	1.000
4	954 240	0.9^272	12 579 840	0.9^387	175 392 000	1.0^5	1.000
5	1 304 576	0.9^236	19 958 400	0.9^369	311 592 960	0.9^5	1.000
6	3 042 816	0.989	44 698 752	0.9^340	711 728 640	0.9^4	1.000
7	3 870 720	0.977	70 785 792	0.9^287	1 193 794 560	0.9^4	1.000
8	6 926 080	0.963	130 032 000	0.9^277	2 393 475 840	0.9^4	1.000
9	8 332 800	0.937	190 834 560	0.9^258	3 784 596 480	0.9^3	1.000
10	15 821 568	0.906	361 525 248	0.9^231	7 444 104 192	0.9^3	1.000
11	14 755 328	0.847	443 931 264	0.988	10 526 745 600	0.9^3	1.000
12	24 487 680	0.792	779 950 080	0.981	19 533 696 000	0.9^3	1.000
13	24 514 560	0.701	1 043 763 840	0.970	27 610 168 320	0.9^287	1.000
14	34 762 240	0.610	1 529 101 440	0.955	47 107 169 280	0.9^279	0.999
15	29 288 448	0.480	1 916 619 264	0.933	64 016 040 960	0.9^266	0.997
16	37 188 480	0.371	2 912 257 152	0.905	107 446 832 640	0.9^247	0.996
17	24 487 680	0.232	3 078 407 808	0.862	134 470 425 600	0.9^217	0.995
18	24 312 960	0.141	4 506 485 760	0.817	218 941 470 720	0.988	0.993
19	10 402 560	0.051	4 946 417 280	0.752	272 302 894 080	0.982	0.988
20	3 230 080	0.012	6 068 256 768	0.680	417 512 148 480	0.974	0.980
21			6 160 876 416	0.592	404 080 834 560	0.962	0.968
22			7 730 384 256	0.502	743 278 970 880	0.948	0.952
23			6 292 581 120	0.389	829 743 344 640	0.927	0.930
24			6 900 969 600	0.298	1 202 317 401 600	0.903	0.905
25			5 479 802 496	0.197	1 334 577 484 800	0.869	0.874
26			4 327 787 520	0.118	1 773 862 272 000	0.831	0.834
27			2 399 241 600	0.055	1 878 824 586 240	0.781	0.786
28			1 197 020 160	0.020	2 496 636 103 680	0.727	0.727
29			163 094 400	0.0^224	2 406 981 104 640	0.656	0.659
30			3 230 080	0.0^45	3 032 021 672 960	0.588	0.583
31					2 841 072 675 840	0.502	0.500
32					3 166 378 709 760	0.421	0.413
33					2 743 311 191 040	0.331	0.325
34					2 877 794 035 200	0.253	0.243
35					2 109 852 702 720	0.171	0.170
36					1 840 136 336 640	0.111	0.109
37					1 109 253 196 800	0.059	0.063
38					689 719 564 800	0.028	0.032
39					230 683 084 800	0.0^279	0.014
40					48 251 508 480	0.0^214	0.005

Appendix Tables 10

Probability function of Σ (for u)

These tables give the distribution of the sum of the number of agreements Σ in paired comparisons of n objects by m observers. Each entry labelled P is the probability that this value of Σ will be attained or exceeded. For $n = 6$, only values of P smaller than the 1 per cent significance level are given.

Table 10A. $m = 3$, $n = 2$ to 8.

$n = 2$		$n = 3$		$n = 4$		$n = 5$		$n = 6$		$n = 7$		$n = 8$	
Σ	P	Σ	P	Σ	P	Σ	P	Σ	P	Σ	P	Σ	P
1	1.000	3	1.000	6	1.000	10	1.000	15	1.000	21	1.000	28	1.000
3	0.250	5	0.578	8	0.822	12	0.944	17	0.987	23	0.998	30	1.000
		7	0.156	10	0.466	14	0.756	19	0.920	25	0.981	32	0.997
		9	0.016	12	0.169	16	0.474	21	0.764	27	0.925	34	0.983
				14	0.038	18	0.224	23	0.539	29	0.808	36	0.945
				16	0.0046	20	0.078	25	0.314	31	0.633	38	0.865
				18	$0.0^3 24$	22	0.020	27	0.148	33	0.433	40	0.736
						24	0.0035	29	0.057	35	0.256	42	0.572
						26	$0.0^3 42$	31	0.017	37	0.130	44	0.400
						28	$0.0^4 30$	33	0.0042	39	0.056	46	0.250
						30	$0.0^6 95$	35	$0.0^3 79$	41	0.021	48	0.138
								37	$0.0^3 12$	43	0.0064	50	0.068
								39	$0.0^4 12$	45	0.0017	52	0.029
								41	$0.0^6 92$	47	$0.0^3 37$	54	0.011
								43	$0.0^7 43$	49	$0.0^4 68$	56	0.0038
								45	$0.0^9 93$	51	$0.0^4 10$	58	0.0011
										53	$0.0^5 12$	60	$0.0^3 29$
										55	$0.0^6 12$	62	$0.0^4 66$
										57	$0.0^8 86$	64	$0.0^4 13$
										59	$0.0^9 44$	66	$0.0^5 22$
										61	$0.0^{10} 15$	68	$0.0^6 32$
										63	$0.0^{12} 23$	70	$0.0^7 40$
												72	$0.0^8 42$
												74	$0.0^9 36$
												76	$0.0^{10} 24$
												78	$0.0^{11} 13$
												80	$0.0^{13} 48$
												82	$0.0^{14} 12$
												84	$0.0^{16} 14$

Table 10B. $m = 4, n = 2$ to 6.

$n = 2$		$n = 3$		$n = 4$		$n = 5$		$n = 5$		$n = 6$		$n = 6$	
Σ	P	Σ	P	Σ	P	Σ	P	Σ	P	Σ	P	Σ	P
2	1.000	6	1.000	12	1.000	20	1.000	42	0.0048	57	0.014	79	$0.0^8 42$
3	0.625	7	0.947	13	0.997	21	1.000	43	0.0030	58	0.0092	80	$0.0^8 28$
6	0.125	8	0.736	14	0.975	22	0.999	44	0.0017	59	0.0058	81	$0.0^9 98$
		9	0.455	15	0.901	23	0.995	45	$0.0^3 73$	60	0.0037	82	$0.0^9 15$
		10	0.330	16	0.769	24	0.979	46	$0.0^3 41$	61	0.0022	83	$0.0^9 12$
		11	0.277	17	0.632	25	0.942	47	$0.0^3 24$	62	0.0013	84	$0.0^{10} 51$
		12	0.137	18	0.524	26	0.882	48	$0.0^4 90$	63	$0.0^3 76$	86	$0.0^{11} 30$
		14	0.043	19	0.410	27	0.805	49	$0.0^4 37$	64	$0.0^3 44$	87	$0.0^{11} 17$
		15	0.025	20	0.278	28	0.719	50	$0.0^4 25$	65	$0.0^3 23$	90	$0.0^{13} 28$
		18	0.0020	21	0.185	29	0.621	51	$0.0^5 93$	66	$0.0^3 13$		
				22	0.137	30	0.514	52	$0.0^5 21$	67	$0.0^4 72$		
				23	0.088	31	0.413	53	$0.0^5 17$	68	$0.0^4 36$		
				24	0.044	32	0.327	54	$0.0^6 74$	69	$0.0^4 18$		
				25	0.027	33	0.249	56	$0.0^7 66$	70	$0.0^5 97$		
				26	0.019	34	0.179	57	$0.0^7 38$	71	$0.0^5 47$		
				27	0.0079	35	0.127	60	$0.0^9 93$	72	$0.0^5 20$		
				28	0.0030	36	0.090			73	$0.0^5 10$		
				29	0.0025	37	0.060			74	$0.0^6 51$		
				30	0.0011	38	0.038			75	$0.0^6 18$		
				32	$0.0^3 16$	39	0.024			76	$0.0^7 78$		
				33	$0.0^4 95$	40	0.016			77	$0.0^7 44$		
				36	$0.0^5 38$	41	0.0088			78	$0.0^7 15$		

Table 10C. $m = 5, n = 2$ to 5.

$n = 2$		$n = 3$		$n = 4$		$n = 5$		$n = 5$	
Σ	P	Σ	P	Σ	P	Σ	P	Σ	P
4	1.000	12	1.000	24	1.000	40	1.000	76	$0.0^4 50$
6	0.375	14	0.756	26	0.940	42	0.991	78	$0.0^4 16$
10	0.063	16	0.390	28	0.762	44	0.945	80	$0.0^5 50$
		18	0.207	30	0.538	46	0.843	82	$0.0^5 15$
		20	0.103	32	0.353	48	0.698	84	$0.0^6 39$
		22	0.030	34	0.208	50	0.537	86	$0.0^6 10$
		24	0.011	36	0.107	52	0.384	88	$0.0^7 23$
		26	0.0039	38	0.053	54	0.254	90	$0.0^8 53$
		30	$0.0^3 24$	40	0.024	56	0.158	92	$0.0^8 12$
				42	0.0093	58	0.092	94	$0.0^9 14$
				44	0.0036	60	0.050	96	$0.0^{10} 46$
				46	0.0012	62	0.026	100	$0.0^{12} 91$
				48	$0.0^3 36$	64	0.012		
				50	$0.0^3 12$	66	0.0057		
				52	$0.0^4 28$	68	0.0025		
				54	$0.0^5 54$	70	0.0010		
				56	$0.0^5 18$	72	$0.0^3 39$		
				60	$0.0^7 60$	74	$0.0^3 14$		

Table 10D. $m = 6, n = 2$ to 4.

$n = 2$		$n = 3$		$n = 4$		$n = 4$		$n = 4$	
Σ	P	Σ	P	Σ	P	Σ	P	Σ	P
6	1.000	18	1.000	36	1.000	55	0.043	74	0.0^412
7	0.688	19	0.969	37	0.999	56	0.029	75	0.0^589
10	0.219	20	0.832	38	0.991	57	0.020	76	0.0^549
15	0.031	21	0.626	39	0.959	58	0.016	77	0.0^532
		22	0.523	40	0.896	59	0.011	80	0.0^668
		23	0.468	41	0.822	60	0.0072	81	0.0^617
		24	0.303	42	0.755	61	0.0049	82	0.0^612
		26	0.180	43	0.669	62	0.0034	85	0.0^734
		27	0.147	44	0.556	63	0.0025	90	0.0^893
		28	0.088	45	0.466	64	0.0016		
		29	0.061	46	0.409	65	0.0^383		
		30	0.040	47	0.337	66	0.0^366		
		31	0.034	48	0.257	67	0.0^348		
		32	0.023	49	0.209	68	0.0^326		
		35	0.0062	50	0.175	69	0.0^316		
		36	0.0029	51	0.133	70	0.0^486		
		37	0.0020	52	0.097	71	0.0^468		
		40	0.0^358	53	0.073	72	0.0^448		
		45	0.0^431	54	0.057	73	0.0^416		

Appendix Table 11

Significance points of $t_{XY.Z}$
(for Kendall's partial rank correlation coefficient)

From Maghsoodloo (1975), Maghsoodloo and Pallos (1981) with permission of the authors and the editor of the *Journal of Computation and Simulation*.

n	0.05	One-tailed level of significance 0.025	0.01	0.005
3	1	1	1	1
4	0.707	1	1	1
5	0.667	0.802	0.816	1
6	0.600	0.667	0.764	0.866
7	0.527	0.617	0.712	0.761
8	0.484	0.565	0.648	0.713
9	0.443	0.515	0.602	0.660
10	0.413	0.480	0.562	0.614
11	0.387	0.453	0.530	0.581
12	0.365	0.430	0.505	0.548
13	0.347	0.410	0.481	0.527
14	0.331	0.391	0.458	0.503
15	0.317	0.375	0.439	0.482
16	0.305	0.361	0.423	0.466
17	0.294	0.348	0.410	0.450
18	0.284	0.336	0.395	0.434
19	0.275	0.326	0.382	0.421
20	0.267	0.317	0.372	0.410
25	0.235	0.278	0.328	0.362
30	0.211	0.251	0.297	0.328

References

Adichie, J. N. (1967a), Asymptotic efficiency of a class of nonparametric tests for regression parameters, *The Annals of Mathematical Statistics*, **38**, 884–93.

Adichie, J. N. (1967b), Estimates of regression parameters based on rank tests, *The Annals of Mathematical Statistics*, **38**, 894–904.

Agresti, A. (1977), Considerations in measuring partial association for ordinal categorical data, *Journal of the American Statistical Association*, **72**, 37–45.

Agresti, A. (1980), Generalized odds ratio for ordinal data, *Biometrics*, **36**, 59–67.

Agresti, A. and J. Pendergast (1986), Comparing mean ranks for repeated measures data, *Communications in Statistics – Theory and Methods*, **15**, 1417–33.

Aiyar, R. J., C. L. Guillier and W. Albers (1979), Asymptotic relative efficiencies of rank tests for trend alternatives, *Journal of the American Statistical Association*, **74**, 226–31.

Alvo, M. and P. Cabilio (1984), A comparison of approximations to the distribution of average Kendall tau, *Communications in Statistics – Theory and Methods*, **13**, 3191–216.

Alvo, M. and P. Cabilio (1985), Average rank correlation statistics in the presence of ties, *Communications in Statistics – Theory and Methods*, **14**, 2095–108.

Alvo, M., P. Cabilio and P. D. Feigin (1982), Asymptotic theory for measures of concordance with special reference to average Kendall tau, *The Annals of Statistics*, **10**, 1269–86.

Beckett, J. and W. R. Schucany (1979), Concordance among categorized groups of judges, *Journal of Educational Statistics*, **4**, 125–37.

Bell, C. B. and K. A. Doksum (1967), Distribution-free tests of independence, *The Annals of Mathematical Statistics*, **38**, 429–46.

Benard, A. and P. Van Elteren (1953), A generalization of the method of m rankings, *Proceedings, Koninklijke Nederlandse Akademie van Wetenschappen, A*, **56**, 358–69, *Indagationes Mathematicae*, **15**.

Bennett, B. M. and M. K. Choe (1984). Studies on tests using Fisher's $\tanh^{-1}$ transformation for Spearman's rho and Kendall's tau in small samples, *Biometrical Journal*, **26**, 631–42.

Best, D. J. (1973), Extended tables for Kendall's tau, *Biometrika*, **60**, 429–30.

Best, D. J. (1974), Tables for Kendall's tau and an examination of the normal approximation. *Technical Paper No. 39*, Division of Mathematical Statistics, Commonwealth Scientific and Industrial Research Organization Australia.

Best, D. J. and P. G. Gipps (1974), The upper tail probabilities of Kendall's tau, *Applied Statistics*, **23**, 98–100.

Best, D. J. and D. E. Roberts (1975), Algorithm AS 89: The upper tail probabilities of Spearman's rho, *Applied Statistics*, **24**, 377–9.

Bhapkar, V. P. (1961), Some nonparametric median procedures, *The Annals of Mathematical Statistics*, **32**, 846–63.

Bhattacharyya, G. K. (1968), Robust estimates of linear trend in multivariate time series, *Annals of the Institute of Statistical Mathematics*, **20**, 299–310.

Bhattacharyya, G. K. (1984), Tests of randomness against trend or serial correlations, in P. R. Krishnaiah and P. K. Sen (eds), *Nonparametric Statistics*, Elsevier Science Publishers, Amsterdam, pp. 89–112.

Bhattacharyya, G. K., R. A. Johnson and H. R. Neave (1970), Percentage points of some non-parametric tests for independence and empirical power comparisons, *Journal of the American Statistical Association*, **65**, 976–83.

Bhuchongkul, S. (1964), A class of nonparametric tests for independence in bivariate populations, *The Annals of Mathematical Statistics*, **35**, 138–49.

Blum, J. R., J. Kiefer and M. Rosenblatt (1961), Distribution free tests of independence based on the sample distribution function, *The Annals of Mathematical Statistics*, **32**, 485–98.

Bobko, P. (1977), A note on Moran's measure of multiple rank correlation, *Psychometrika*, **42**, 311–14.

Bose, R. C. (1956), Paired comparison designs for testing concordance between judges, *Biometrika*, **43**, 113–21.

Bradley, R. A. (1954), The rank analysis of incomplete block designs, II. Additional tables for the method of paired comparisons, *Biometrika*, **41**, 502–37; Corrigenda, *Ibid.*, **51** (1964), 288–98.

Bradley, R. A. (1955), Rank analysis of incomplete block designs. III. Some large sample results on estimation and power for a method of paired comparisons, *Biometrika*, **42**, 450–70.

Bradley, R. A. (1976), Science, statistics and paired comparisons, *Biometrics*, **32**, 213–32.

Bradley, R. A. (1984), Paired comparisons: Some basic procedures and examples, in P. R. Krishnaiah and R. K. Sen (eds), *Nonparametric Statistics*, Elsevier Science Publishers, Amsterdam, pp. 299–326.

Bradley, R. A. and M. E. Terry (1952), The rank analysis of incomplete blocks designs, I. The method of paired comparisons, *Biometrika*, **39**, 324–45.

Bright, H. F. (1954), A method for computing the Kendall tau coefficient, *Educational and Psychological Measurement*, **14**, 700–4.

Brits, S. J. M. and H. H. Lemmer (1986), Nonparametric tests for treatment effects after a preliminary test on block effects in a randomized block design, *South African Journal of Statistics*, **20**, 45–65.

Brown, B. W., M. Hollander and R. M. Korwar (1974), Nonparametric tests of independence for censored data, with applications to heart transplant studies, in F. Proschan and R. J. Serfling (eds), *Reliability and Biometry: Statistical Analysis of Lifelength*, Society for Industrial and Applied Mathematics, Philadelphia, PA, pp. 327–54.

Brunden, M. N. and N. R. Mohberg (1976), The Benard–van Elteren statistic and nonparametric computation, *Communications in Statistics – Simulation and Computation*, **5**, 155–62.

Buck, W. (1980), Tests of significance for point-biserial rank correlation coefficients in the presence of ties, *Biometrical Journal*, **22**, 153–8.

Buckley, M. J. and G. K. Eagleson (1986), Assessing large sets of rank correlations, *Biometrika*, **73**, 151–7.

Burnett, R. T. and A. R. Willan (1988), Linear rank tests for randomized block designs, *Communications in Statistics – Theory and Methods*, **17**, 2455–70.

Burr, E. J. (1960), The distribution of Kendall's score *S* for a pair of tied rankings, *Biometrika*, **47**, 151–71.

Chambers, E. G. (1946), Statistical techniques in applied psychology. *Biometrika*, **33**, 269–73.

Cochran, W. G. and G. M. Cox (1957), *Experimental Designs*, John Wiley & Sons, New York.

Cohen, A. (1982), Analysis of large sets of ranking data, *Communications in Statistics –Theory and Methods*, **11**, 235–56.

Conover, W. J. and R. L. Iman (1976), On some alternative procedures using ranks for the analysis of experimental designs, *Communications in Statistics – Theory and Methods*, **5**, 1349–68.

Costello, P. S. and D. A. Wolfe (1985), A new nonparametric approach to the problem of agreement between two group of judges, *Communications in Statistics – Simulation and Computation*, **14**, 791–805.

Cox, D. R. and A. Stuart (1955), Some quick sign tests for trend in location and dispersion, *Biometrika*, **42**, 80–95.

Cureton, E. E. (1958), The average correlation coefficient when ties are present, *Psychometrika*, **23**, 271–2; Correction, *Ibid.*, **30**, 377.

Daniels, H. E. (1944), The relation between measures of correlation in the universe of sample permutations, *Biometrika*, **33**, 129–35.

Daniels, H. E. (1948), A property of rank correlations, *Biometrika*, **35**, 416–47.

Daniels, H. E. (1950), Rank correlation and population models, *Journal of the Royal Statistical Society, B*, **12**, 171–81.

Daniels, H. E. (1951), Note on Durbin and Stuart's formula for $E(r_s)$, *Journal of the Royal Statistical Society, B*, **13**, 310.

Daniels, H. E. (1954), A distribution-free test for regression parameters, *The Annals of Mathematical Statistics*, **25**, 499–513.

Daniels, H. E. and M. G. Kendall (1947), The significance of rank correlations where parental correlation exists, *Biometrika*, **34**, 197–208.

Dantzig, G. B. (1939), On a class of distributions that approach the normal distribution function, *The Annals of Mathematical Statistics*, **10**, 247–53.

David, F. N. and E. Fix (1961), Rank correlation and regression in a non normal surface, in *Proceedings of the Fourth Berkeley Symposium on Mathematical Statistics and Probability*, Vol. I, University of California Press, Berkeley, pp. 177–97.

David, F. N. and C. L. Mallows (1961), The variance of Spearman's rho in normal samples, *Biometrika*, **48**, 19–28.

David, H. A. (1988), *The Method of Paired Comparisons* (2nd edn), Charles Griffin & Company, London.

David, S. T., M. G. Kendall and A. Stuart (1951), Some questions of distribution in the theory of rank correlation, *Biometrika*, **38**, 131–40.

Davidson, R. R. and P. H. Farquahr (1976), A bibliography on the method of paired comparisons, *Biometrika*, **32**, 241–52.

Davis, J. A. (1967), A partial coefficient for Goodman and Kruskal's gamma, *Journal of the American Statistical Association*, **62**, 189–93.

De Jonge, C. and M. A. J. Van Montfort (1972), The null distribution of Spearman's S when $n = 12$, *Statistica Neerlandica*, **26**(1), 15–17.

De Kroon, J. and P. Van der Laan (1983), A generalization of Friedman's rank statistic, *Statistica Neerlandica*, **37**, 1–14.

Diaconis, P. and R. L. Graham (1977), Spearman's footrule as a measure of disarray, *Journal of the Royal Statistical Society, B*, **39**, 262–8.

Dietz, E. J. (1987), A comparison of robust estimators in simple regression, *Communications in Statistics – Simulation and Computation*, **16**, 1209–27.

Dietz, E. J. (1989), Teaching regression in a nonparametric statistics course, *The American Statistician*, **43**, 35–40.

Dietz, E. J. and T. J. Killeen (1981), A nonparametric multivariate test for monotone trend with pharmaceutical applications, *Journal of the American Statistical Association*, **76**, 169–74.

Dixon, W. J. (1953), Power function of the sign test and power efficiency for normal alternatives, *The Annals of Mathematical Statistics*, **24**, 467–73.

Downton, F. (1976), Nonparametric tests for block experiments, *Biometrika*, **63**, 137–41.

Dubois, P. (1939), Formulas and tables for rank correlation, *Psychological Record*, **3**, 46–56.

Dunstan, F. D. J., A. B. J. Nix and J. F. Reynolds (1979), *Statistical Tables*, RND Publications, Cardiff.

Durbin, J. (1951), Incomplete blocks in ranking experiments, *British Journal of Psychology* (Statistical Section), **4**, 85–90.

Durbin, J. and A. Stuart (1951), Inversions and rank correlation coefficients, *Journal of the Royal Statistical Society, B*, **13**, 303–9.

Dykstra, O. (1956), A note on the rank analysis of incomplete block designs; Applications beyond the scope of existing tables, *Biometrics*, **12**, 301–6.

Eells, W. C. (1929), Formulas for probable errors of coefficients of correlation, *Journal of the American Statistical Association*, **24**, 170–3.

Ehrenberg, A. S. C. (1952), On sampling from a population of rankers, *Biometrika*, **39**, 82–7.

Esscher, F. (1924), On a method of determining correlation from the ranks of variates, *Skandinavisk Aktuarietidskrift*, **7**, 201–19.

Evans, L. S. (1973), A mechanical interpretation of the coefficient of rank correlation and other analogies, *The American Statistician*, **27**, 79–81.

Farlie, D. J. G. (1960), The performance of some correlation coefficients for a general bivariate distribution, *Biometrika*, **47**, 307–23.

Farlie, D. J. G. (1961), The asymptotic efficiency of Daniels's generalized correlation coefficient, *Journal of the Royal Statistical Society, B*, **23**, 128–42.

Farlie, D. J. G. (1963), The asymptotic efficiency of Daniels's generalized correlation coefficient, *Biometrika*, **50**, 499–504.

Fawcett, R. F. and K. C. Salter (1984), A Monte Carlo study of the *F*-test and three tests based on ranks of treatment effects in randomized block designs, *Communications in Statistics – Simulation and Computation*, **13**, 213–25.

Fawcett, R. F. and K. C. Salter (1987), Distributional studies and the computer: An analysis of Durbin's rank test, *The American Statistician*, **41**, 81–3.

Feigin, P. D. and M. Alvo (1986), Intergroup diversity and concordance for ranking data: An approach via metrics for permutations, *The Annals of Statistics*, **14**, 691–707.

Feigin, P. D. and A. Cohen (1978), On a model of concordance between judges, *Journal of the Royal Statistical Society, B*, **40**, 203–13.

Feller, W. (1945), The fundamental limit theorems in probability, *Bulletin of the American Mathematical Society*, **51**, 800–32.

Ferretti, N. E. (1987), Efficiency of tests based on weighted rankings for randomized blocks when the alternatives have an *a priori* ordering, *Communications in Statistics – Theory and Methods*, **16**, 1629–53.

Ferretti, N. E. and V. J. Yohai (1986), Efficiency of tests based on weighted rankings for equality of treatment effects in a complete randomized blocks layout, *Communications in Statistics – Theory and Methods*, **15**, 1179–200.

Fieller, E. C., H. O. Hartley and E. S. Pearson (1957), Tests for rank correlation coefficients: I, *Biometrika*, **44**, 470–81.

Fieller, E. C. and E. S. Pearson (1961), Tests for rank correlation coefficients: II, *Biometrika*, **48**, 29–40.

Fisher, R. A. (1934), *Statistical Methods for Research Workers*, Oliver & Boyd, Edinburgh.

Fligner, M. A. and J. S. Verducci (1987), Aspects of two group concordance, *Communications in Statistics – Theory and Methods*, **16**, 1479–503.

Franklin, L. A. (1987a), Approximations, convergence and exact tables for Spearman's rank correlation coefficient, *Proceedings of the Statistical Computing Section*, American Statistical Association, August, 244–7.

Franklin, L. A. (1987b), The complete exact null distribution of Spearman's rho for $n = 12(1)16$, *Proceedings of the 19th Symposium on the Interface between Computer Science and Statistics*, American Statistical Association, March, 337–42.

Franklin, L. A. (1988a), A note on approximations and convergence in distribution for Spearman's rank correlation coefficient, *Communications in Statistics – Theory and Methods*, **17**, 55–9.

Franklin, L. A. (1988b), Exact tables of Spearman's footrule for $n = 11 (1) 18$ with estimate of convergence and errors for the normal approximation, *Statistics & Probability Letters*, **6**, 399–406.

Franklin, L. A. (1988c), The complete exact null distribution of Spearman's rho for $n = 12(1)18$, *Journal of Statistical Computation and Simulation*, **29**(3), 255–269.

Franklin, L. A. (1989), A note on the Edgeworth approximation to the distribution of Spearman's rho with a correction to Pearson's approximation, *Communications in Statistics – Simulation and Computation*, **18**, 245–52.

Friedman, J. H. and L. Rafsky (1983), Graph-theoretic measures of multi-variate association and prediction, *The Annals of Statistics*, **11**, 377–91.

Friedman, M. (1937), The use of ranks to avoid the assumption of normality implicit in the analysis of variance, *Journal of the American Statistical Association*, **32**, 675–701.

Friedman, M. (1940), A comparison of alternative tests of significance for the problem of *m* rankings, *The Annals of Mathematical Statistics*, **11**, 86–92.

Gerig, T. M. (1969), A multivariate extension of Friedman's χ_r^2 test, *Journal of the American Statistical Association*, **64**, 1595–608.

Gerig, T. M. (1975), A multivariate extension of Friedman's χ_r^2 test with random covariates, *Journal of the American Statistical Association*, **70**, 443–7.

Ghosh, M. (1975), On some properties of a class of Spearman rank statistics with applications, *Annals of the Institute of Statistical Mathematics, Tokyo*, **27**, 57–68.

Gilbert, R. O. (1972), A Monte Carlo study of analysis of variance and competing rank tests for Scheffe's mixed model, *Journal of the American Statistical Association*, **67**, 71–5.

Glasser, G. J. and R. F. Winter (1961), Critical values of the coefficient of rank correlation for testing the hypothesis of independence, *Biometrika*, **48**, 444–8.

Gokhale, D. V. (1968), On asymptotic relative efficiencies of a class of rank tests of independence of two varieties, *Annals of the Institute of Statistical Mathematics*, **20**, 255–61.

Goodman, L. A. (1959), Partial tests for partial taus, *Biometrika*, **46**, 425–32.

Goodman, L. A. (1984), *The Analysis of Cross-classified Data Having Ordered Categories*, Harvard University Press, Cambridge, MA.

Goodman, L. A. and Y. Grunfeld (1961), Some nonparametric tests for comovements between time series, *Journal of the American Statistical Association*, **56**, 11–26.

Goodman, L. A. and W. H. Kruskal (1954), Measures of association for cross classifications, *Journal of the American Statistical Association*, **49**, 732–64; Correction, *Ibid.*, **52**, 578.

Goodman, L. A. and W. H. Kruskal (1959), Measures of association for cross classifications. II: Further discussion and references, *Journal of the American Statistical Association*, **54**, 123–63.

Goodman, L. A. and W. H. Kruskal (1963), Measures of association for cross classifications. III: Approximate sampling theory, *Journal of the American Statistical Association*, **58**, 310–64.

Goodman, L. A. and W. H. Kruskal (1972), Measures of association for cross classifications. IV: Simplification of asymptotic variances, *Journal of the American Statistical Association*, **67**, 415–21.

Goodman, L. A. and W. H. Kruskal (1979), *Measures of Association for Cross Classicifications*, Springer-Verlag, New York.

Gordon, A. D. (1979a), A measure of the agreement between rankings, *Biometrika*, **66**, 7–15.

Gordon, A. D. (1979b), Another measure of the agreement between rankings, *Biometrika*, **66**, 327–32.

Govindarajulu, A. (1976), Asymptotic normality and efficiency of a class of test statistics, in *Essays in Probability and Statistics*, Shinko Tsusho, Tokyo, pp. 535–58.

Greiner, R. (1909), Über das Fehlersystem der Kollektivmasslehre, *Zeitschift für Mathematik und Physik*, **57**, 121–58, 225–60, 337–73.

Griffin, H. D. (1958), Graphic computation of tau as a coefficient of disarray, *Journal of the American Statistical Association*, **53**, 441–7.

Griffiths, D. (1980), A pragmatic approach to Spearman's rank correlation coefficient, *Teaching Statistics*, **2**, 10–13.

Groggel, D. J. (1987), A Monte Carlo study of rank tests for block designs, *Communications in Statistics – Simulation and Computation*, **16**, 601–20.

Groggel, D. J. and J. H. Skillings (1986), Distribution-free tests for main effects in multifactor designs, *The American Statistician*, **40**, 99–102.

Guttman, L. (1946), An approach for quantifying paired comparisons and rank order, *The Annals of Mathematical Statistics*, **17**, 144–63.

Haden, H. G. (1947), A note on the distribution of the different orderings of *n* objects, *Proceedings of the Cambridge Philosophical Society*, **43**, 1–9.

Harter, H. L. (1961), Expected values of normal order statistics, *Biometrika*, **48**, 151–65.

Hays, W. L. (1960), A note on average tau as a measure of concordance, *Journal of the American Statistical Association*, **55**, 331–41.

Henery, R. J. (1986), Interpretation of average ranks, *Biometrika*, **73**, 224–7.

Henze, F. H.-H. (1979), The exact noncentral distributions of Spearman's *r* and other related correlation coefficients, *Journal of the American Statistical Association*, **74**, 459–64.

Hoeffding, W. (1947), On the distribution of the rank correlation coefficient when the variates are not independent, *Biometrika*, **34**, 183–96.

Hoeffding, W. (1948a), A class of statistics with asymptotically normal distribution, *The Annals of Mathematical Statistics*, **19**, 293–325.

Hoeffding, W. (1948b), A non-parametric test of independence, *The Annals of Mathematical Statistics*, **19**, 546–57.

Hoeffding, W. (1951), Optimum, non-parametric tests, *Proceedings of the Second Berkeley Symposium on Mathematical Statistics and Probability*, University of California Press, Berkeley, pp. 83–92.

Hoeffding, W. (1952), The large sample power of tests based on permutations of observations, *The Annals of Mathematical Statistics*, **23**, 169–72.

Hoflund, O. (1963), Simulated distributions for small *n* of Kendall's partial rank correlation coefficient, *Biometrika*, **50**, 520–1.

Hollander, M. (1967), Rank tests for randomized blocks when the alternatives have a prior ordering, *The Annals of Mathematical Statistics*, **38**, 867–77.

Hollander, M. and J. Sethuraman (1978), Testing for agreement between two groups of judges, *Biometrika*, **65**, 403–11.

Hora S. C. and R. L. Iman (1988), Asymptotic relative efficiencies of the rank-transformation procedure in randomized complete block designs, *Journal of the American Statistical Association*, **83**, 462–70.

Hotelling, H. and M. R. Pabst (1936), Rank correlation and tests of significance involving no assumptions of normality, *The Annals of Mathematical Statistics*, **7**, 29–43.

Hubert, L. J. (1979), Generalized concordance, *Psychometrika*, **44**, 135–42.

Hussein, S. S. and P. Sprent (1983), Non-parametric regression, *Journal of the Royal Statistical Society, A*, **146**, 182–191.

Hutchinson, T. P. (1976), Combining two-tailed rank-correlation statistics, *Applied Statistics*, **25**, 21–5.

Iman, R. L. and W. J. Conover (1978), Approximations of the critical region for Spearman's rho with and without ties present. *Communications in Statistics – Simulation and Computation*, **7**, 269–82.

Iman, R. L. and J. M. Davenport (1980), Approximations of the critical region of the Friedman statistic, *Communications in Statistics – Theory and Methods*, **9**, 571–95.

Iman, R. L., S. C. Hora and W. J. Conover (1984), Comparison of asymptotically distribution-free procedures for the analysis of complete blocks, *Journal of the American Statistical Association*, **79**, 674–85.

Irwin, J. O. (1925), The further theory of Francis Galton's individual difference problem, *Biometrika*, **17**, 100–28.

Jensen, D. R. (1974), On the joint distribution of Friedman's χ_r^2 statistic, *The Annals of Statistics*, **2**, 311–22.

Jensen, D. R. (1977), On approximating the distributions of Friedman's χ_r^2 and related statistics, *Metrika*, **24**, 75–85.

Jirina, M. (1976), On the asymptotic normality of Kendall's rank correlation statistic, *The Annals of Statistics*, **4**, 214–15.

Joag-Dev, K. (1984), Measures of dependence, in P. R. Krishnaiah and P. K. Sen (eds), *Nonparametric Statistics*, Elsevier Science Publishers, Amsterdam, pp. 937–58.

Johnson, N. S. (1979), Nonnull properties of Kendall's partial rank correlation coefficient, *Biometrika*, **66**, 333–7.

Jonckheere, A. R. (1954a), A distribution-free k-sample test against ordered alternatives, *Biometrika*, **41**, 133–45.

Jonckheere, A. R. (1954b), A test of significance for the relation between m rankings and k ranked categories, *British Journal of Statistical Psychology*, **7**, 93–100.

Kaarsemaker, L. and A. Van Wijngaarden (1952), Tables for use in rank correlation, *Report R 73*, Computation Mathematical Centre, Amsterdam.

Kaarsemaker, L. and A. Van Wijngaarden (1953), Tables for use in rank correlation, *Statistica Neerlandica*, **7**, 41–54.

Katz, B. M. and M. McSweeney (1983), Some non-parametric tests for analyzing ranked data in multi-group repeated measures designs, *British Journal of Mathematical and Statistical Psychology*, **36**, 145–56.

Kendall, M. G. (1938), A new measure of rank correlation, *Biometrika*, **30**, 91–3.

Kendall, M. G. (1942a), Note on the estimation of a ranking, *Journal of the Royal Statistical Society, A*, **105**, 119–21.

Kendall, M. G. (1942b), Partial rank correlation, *Biometrika*, **32**, 277–83.

Kendall, M. G. (1945), The treatment of ties in ranking problems, *Biometrika*, **33**, 239–51.

Kendall, M. G. (1947), The variance of τ when both rankings contain ties, *Biometrika*, **34**, 297–8.

Kendall, M. G. (1949), Rank and product-moment correlation, *Biometrika*, **36**, 177–93.

Kendall, M. G. (1955), Further contributions to the theory of paired comparisons, *Biometrics*, **11**, 43–62.

Kendall, M. G., S. F. H. Kendall and B. B. Smith (1939), The distribution of Spearman's coefficient of rank correlation in a universe in which all rankings occur an equal number of times, *Biometrika*, **30**, 251–73.

Kendall, M. G. and B. B. Smith (1939), The problem of *m* rankings, *The Annals of Mathematical Statistics*, **10**, 275–87.

Kendall, M. G. and B. B. Smith (1940), On the method of paired comparisons, *Biometrika*, **31**, 324–45.

Kendall, M. G. and A. Stuart (1958), *The Advanced Theory of Statistics*, Vol. 2, Charles Griffin & Company, London.

Kendall, M. G. and A. Stuart (1979), *The Advanced Theory of Statistics*, Vol. 2, Charles Griffin & Company, London.

Kepner, J. L. and D. H. Robinson (1984), A distribution-free rank test for ordered alternatives in randomized complete block designs, *Journal of the American Statistical Association*, **79**, 212–17.

Kerridge, D. (1975), The interpretation of rank correlations, *Applied Statistics*, **24**, 257–8.

Kirk, R. E. (1968), *Experimental Design: Procedures for the Behavioral Sciences*, Brooks/Cole Publishing Company, Belmont, CA.

Knight, W. R. (1966), A computer method for calculating Kendall's tau with ungrouped data, *Journal of the American Statistical Association*, **61**, 436–9.

Kochar, S. C. and R. P. Gupta (1987), Competitors of the Kendall-tau test for testing independence against positive quadrant dependence, *Biometrika*, **74**, 664–6.

Konijn, H. S. (1956), On the power of certain tests for independence in bivariate populations, *The Annals of Mathematical Statistics*, **27**, 300–23; Correction, *Ibid.*, **29**, (1958), 935–6.

Konijn, H. S. (1961), Non-parametric, robust and short-cut methods in regression and structural analysis, *The Australian Journal of Statistics*, **3**, 77–86.

Korn, E. L. (1984), Kendall's tau with a blocking variable, *Biometrics*, **40**, 209–14.

Koroljuk, V. S. and J. V. Borovskih (1982), Approximation to the distribution of Spearman's rank correlation coefficient, *Theory of Probability and Mathematical Statistics*, **25**, 51–9.

Kraemer, H. C. (1974), The non-null distribution of the Spearman rank correlation coefficient, *Journal of the American Statistical Association*, **69**, 114–17.

Kraemer, H. C. (1976), The small sample non-null properties of Kendall's coefficient of concordance for normal populations, *Journal of the American Statistical Association*, **71**, 608–13.

Kraemer, H. C. (1981), Intergroup concordance: Definition and estimation, *Biometrika*, **68**, 641–6.

Kraemer, H. C. (1985), A strategy to teach the concept and application of power of statistical tests, *Journal of Educational Statistics*, **10**, 173–95.

Kruskal, W. H. (1952), A non-parametric test for the several sample problem, *The Annals of Mathematical Statistics*, **23**, 525–40.

Kruskal, W. H. (1958), Ordinal measures of association, *Journal of the American Statistical Association*, **53**, 814-61.

Kruskal, W. H. and W. A. Wallis (1952), Use of ranks in one-criterion variance-analysis, *Journal of the American Statistical Association*, **47**, 583-621; Correction, *Ibid.*, **48**, 907-11.

Lancaster, J. F. and D. Quade (1984), A nonparametric test for linear regression based on combining Kendall's tau with the sign test, *Journal of the American Statistical Association*, **80**, 393-7.

Lehmann, E. L. (1951), Consistency and unbiasedness of certain nonparametric tests, *The Annals of Mathematical Statistics*, **22**, 165-79.

Lehmann, E. L. (1953), The power of rank tests, *The Annals of Mathematical Statistics*, **24**, 23-43.

Lehmann, E. L. (1966), Some concepts of dependence, *The Annals of Mathematical Statistics*, **37**, 1137-53.

Lehmann, R. (1977), General derivation of partial and multiple rank correlation coefficients, *Biometrical Journal*, **19**, 229-36.

Lemmer, H. H., D. J. Stoker and S. G. Reinach (1968), A distribution-free analysis of variance technique for block designs, *South African Journal of Statistics*, **2**, 9-32.

Li, L. and W. R. Schucany (1975), Some properties of a test for concordance of two groups of rankings, *Biometrika*, **62**, 417-23.

Lieberson, S. (1961), Non-graphic computation of Kendall's tau, *The American Statistician*, **15**(4), 20-1.

Lindeberg, J. W. (1925), Über die Korrelation, *VI Skandinavisk Matematiker Kongress Kobenhavn*, **1**, 437-46.

Lindeberg, J. W. (1929), Some remarks on the mean error rate of the percentage of correlation, *Nordisk Statistik Tidskrift*, **1**, 137-41.

Linhart, H. (1960), Approximate test for *m* rankings, *Biometrika*, **47**, 476-80.

Lyerly, S. B. (1952), The average Spearman rank correlation coefficient, *Psychometrika*, **17**, 421-28.

Mack, G. A. and J. H. Skillings (1980), A Friedman-type rank test for main effects in a two-factor ANOVA, *Journal of American Statistical Association*, **75**, 947-51.

Maghsoodloo, S. (1975), Estimates of the quantiles of Kendall's partial rank correlation coefficient and additional quantile estimates, *Journal of Statistical Computation and Simulation*, **4**, 155-64.

Maghsoodloo, S. and L. L. Pallos (1981), Asymptotic behavior of Kendall's partial rank correlation coefficient and additional quantile estimates, *Journal of Statistical Computation and Simulation*, **13**, 41-8.

Mallows, C. L. (1957), Non-null ranking models I, *Biometrika*, **44**, 114-30.

Mann, H. B. (1945), Non-parametric tests against trend, *Econometrica*, **13**, 245-59.

Mann, H. B. and D. R. Whitney (1947), On a test of whether one of two random variables is larger than the other, *The Annals of Mathematical Statistics*, **18**, 50-60.

Mardia, K. V. (1969), The performance of some tests for independence for contingency-type bivariate distributions, *Biometrika*, **56**, 449-51.

Markowski, E. P. (1984), Inference using near-neighbour trimmed rank statistics for simple linear regression models, *Biometrika*, **71**, 51-6.

Markowski, E. P. (1987), Comparing tests judged asymptotically equally efficient, *Communications in Statistics – Simulation and Computation*, **16**, 629–43.

Maritz, J. S. (1979), On Theil's method of distribution-free regression, *Australian Journal of Statistics*, **21**, 30–5.

Mehra, K. L. and J. Sarangi (1967), Asymptotic efficiency of certain rank tests for comparative experiments, *The Annals of Mathematical Statistics*, **38**, 90–107.

Michaelis, J. (1971), Schwellenwerte des Friedman-tests, *Biometrische Zeitschrift*, **13**, 118–29.

Monjardet, B. and C. LeConte de Poly-Barbut (1986), Valeurs extrémales de la différence des deux coefficients de correlation de rangs rho et tau, *Comptes Rendus des Séances de l'Académie des Sciences*, Serie 1, Mathématique, **303**(10), 483–6.

Moore, G. H. and W. A. Wallis (1943), Time series significance tests based on signs of differences, *Journal of the American Statistical Association*, **38**, 153–64.

Moran, P. A. P. (1947), On the method of paired comparisons, *Biometrika*, **34**, 363–5.

Moran, P. A. P. (1948a), Rank correlation and permutation distributions, *Proceedings of the Cambridge Philosophical Society*, **44**, 142–4.

Moran, P. A. P. (1948b), Rank correlation and product-moment correlation, *Biometrika*, **35**, 203–6.

Moran, P. A. P. (1950a), A curvilinear ranking test, *Journal of the Royal Statistical Society, B*, **12**, 292–5.

Moran, P. A. P. (1950b), Recent developments in ranking theory, *Journal of the Royal Statistical Society, B*, **12**, 153–62.

Moran, P. A. P. (1951), Partial and multiple rank correlation, *Biometrika*, **38**, 26–32.

Moran, P. A. P. (1979), The use of correlation in large samples, *The Australian Journal of Statistics*, **21**, 293–300.

Mosteller, F. (1951a,b,c), Remarks on the method of paired comparisons, *Psychometrika*, **16**, 3–9, 203–6, 207–18.

Muhsam, H. V. (1954), A probability approach to ties in rank correlation, *Bulletin of the Research Council of Israel*, **3**, 321–7.

Neave, H. R. (1981), *Elementary Statistics Tables*, George Allen and Unwin, London, England.

Nelson, L. S. (1986), Critical values for sums of squared rank differences in Spearman's correlation test, *Journal of Quality Technology*, **18**, 194–6.

Nijsse, M. (1988), Testing the significance of Kendall's τ and Spearman's r_s, *Psychological Bulletin*, **103**, 235–7.

Noether, G. E. (1958), The efficiency of some distribution-free tests, *Statistica Neerlandica*, **12**, 63–73.

Noether, G. E. (1967), *Elements of Nonparametric Statistics*, John Wiley, New York.

Noether, G. E. (1981), Why Kendall tau?, *Teaching Statistics*, **3**, 41–3.

Noether, G. E. (1985), Elementary estimates: An introduction to nonparametrics, *Journal of Educational Statistics*, **10**, 211–22.

Noether, G. E. (1987a), Sample size determination for Kruskal–Wallis and Friedman tests, *Proceedings of the International Statistical Institute*, Tokyo.

Noether, G. E. (1987b), Sample size determination for some common non-parametric tests, *Journal of the American Statistical Association*, **82**, 645–7.

Odeh, R. E. (1977), Extended tables of the distribution of Friedman's S-statistic in the two-way layout, *Communications in Statistics – Simulation and Computation*, **B6**, 29–48.

Olds, E. G. (1938), Distribution of sums of squares of rank differences for small numbers of individuals, *The Annals of Mathematical Statistics*, **9**, 133–48.

Olds, E. G. (1949), The 5% significance levels for sums of squares of rank differences and a correction, *The Annals of Mathematical Statistics*, **20**, 117–18.

Otten, A. (1973a), Note on the Spearman rank correlation coefficient, *Journal of the American Statistical Association*, **68**, 585.

Otten, A. (1973b), The null distribution of Spearman's S when $n = 13(1)16$, *Statistica Neerlandica*, **27**(1), 19–20.

Owen, D. B. (1962), *Handbook of Statistical Tables*, Addison-Wesley, Reading, MA.

Page, E. B. (1963), Ordered hypotheses for multiple treatments: A significance test for linear ranks, *Journal of the American Statistical Association*, **58**, 216–30.

Palachek, A. D. and R. A. Kerin (1982), Alternative approaches to the two group concordance problem in brand preference rankings, *Journal of Marketing Research*, **19**, 386–9.

Palachek, A. D. and W. R. Schucany (1983), On the correlation of a group of rankings with an external ordering relative to the internal concordance, *Statistics and Probability Letters*, **1**, 259–63.

Palachek, A. D. and W. R. Schucany (1984), On approximate confidence intervals for measures of concordance, *Psychometrika*, **49**, 133–41.

Papaioannou, T. and S. Loukas (1984), Inequalities on rank correlation with missing data, *Journal of the Royal Statistical Society, B*, **46**, 68–71.

Papaioannou, T. and T. Speevak (1977), Rank correlation inequalities with missing data, *Communications in Statistics – Theory and Methods, A*, **6**, 67–72.

Patel, K. M. (1975), A generalized Friedman test for randomized block designs when observations are subject to right censorship, *Communications in Statistics – Theory and Methods*, **4**, 389–94.

Paterson, L. J. (1988), Some recent work on making incomplete-block designs available as a tool for science, *International Statistical Review*, **56**, 129–38.

Pearson, K. (1907), Mathematical contributions to the theory of evolution. XVI. On further methods of determining correlation, *Drapers' Co. Research Memoirs*, Biometric Series IV, Cambridge University Press.

Pearson, K. (1914), On an extension of the method of correlation by grades or ranks, *Biometrika*, **10**, 416–18.

Pearson, K. (1921), Second note on the coefficient of correlation as determined from the quantitative measurement of one variate and the ranking of a second variate, *Biometrika*, **13**, 302–5.

Pearson, K. and M. V. Pearson (1931), On the mean character and variance of a ranked individual and on the mean and variance of the intervals between ranked individuals, *Biometrika*, **23**, 364–97; Addition, *Ibid.*, **24**, 203.

Pearson, K. and B. A. S. Snow (1962), Tests for rank correlation coefficients. III: Distribution of the transformed Kendall coefficient, *Biometrika*, **49**, 185–92.

Pitman, E. J. G. (1937a), Significance tests which may be applied to samples from any populations, *Journal of the Royal Statistical Society, B*, **4**, 119–30.

Pitman, E. J. G. (1937b), Significance tests which may be applied to samples from any populations, II: The correlation coefficient test, *Journal of the Royal Statistical Society, B*, **4**, 225–32.

Pitman, E. J. G. (1938), Significance tests which may be applied to samples from any populations, III: The analyses of variance tests, *Biometrika*, **29**, 322–35.

Prentice, M. J. (1979), On the problem of m incomplete rankings, *Biometrika*, **65**, 167–9.

Puri, M. L. and P. K. Sen (1971), *Nonparametric Methods in Multivariate Analysis*, John Wiley, New York.

Quade, D. (1972), Average internal rank correlation, *Technical Report SW 16/72*, Mathematische Centrum, Amsterdam.

Quade, D. (1974), Nonparametric partial correlation, in H. M. Blalock, Jr. (ed), *Measurement in the Social Sciences*, Aldine, Chicago, pp. 369–98.

Quade, D. (1979), Using weighted rankings in the analysis of complete blocks with additive block effects, *Journal of the American Statistical Association*, **74**, 680–3.

Quade, D. (1984), Nonparametric methods in two-way layouts, in P. R. Krishnaiah and P. K. Sen (eds), *Nonparametric Statistics*, Elsevier Science Publishers, Amsterdam, pp. 185–228.

Ramsay, P. H. (1989), Critical values for Spearman's rank order correlation, *Journal of Educational Statistics*, in press.

Randles, R. H. (1988), Theil test and estimator of slope, in S. Kotz and N. L. Johnson (eds), *Encyclopedia of Statistical Sciences*, **9**, John Wiley, NY, pp. 226–7.

Reynolds, H. T. (1974), Ordinal partial correlation and causal inferences, in H. M. Blalock, Jr. (ed), *Measurement in the Social Sciences*, Aldine, Chicago, pp. 399–423.

Robillard, P. (1972), Kendall's S distribution with ties in one ranking, *Journal of the American Statistical Association*, **67**, 453–5.

Rosander, A. C. (1942), The use of inversions as a test of random order, *Journal of the American Statistical Association*, **37**, 352–8.

Rosenthal, I. and T. S. Ferguson (1965), An asymptotic distribution-free multiple comparison method with applications to the problem of n rankings of m objects, *British Journal of Mathematical Psychology*, **18**, 243–254.

Ruymgaart, F. H. (1973), Asymptotic theory of rank tests of independence, *Technical Report 43*, Mathematische Centrum, Amsterdam.

Ryans, A. B. (1976), Evaluating aggregated predictions from models of consumer choice behavior, *Journal of Marketing Research*, **13**, 333–8.

Ryans, A. B. and V. Srinivasan (1979), Improved method for comparing rank-order preferences of two groups of consumers, *Journal of Marketing Research*, **16**, 583–7.

Salama, I. A. and D. Quade (1981), Using weighted rankings to test against ordered alternatives in complete blocks, *Communications in Statistics – Theory and Methods*, **10**, 385–99.

Salama, I. A. and D. Quade (1982), A nonparametric comparison of two multiple regressions by means of a weighted measure of correlation, *Communications in Statistics – Theory and Methods*, **11**, 1185–95.

Salter, K. C. and R. F. Fawcett (1985), A robust and powerful rank test of treatment effects in balanced incomplete block designs, *Communications in Statistics –Simulation and Computation*, **14**, 807–28.

Savage, I. R. (1952), *Bibliography of Nonparametric Statistics*, Harvard University Press, Cambridge, MA.

Savage, I. R. (1954), *Contributions to the theory of rank order statistics*, Memorandum for private circulation.

Schemper, M. (1984a), A generalized Friedman test for data defined by intervals, *Biometrical Journal*, **26**, 305–8.

Schemper, M. (1984b), Exact test procedures for generalized Kendall correlation coefficients, *Biometrical Journal*, **26**, 399–406.

Schemper, M. (1987), One- and two-sample tests of Kendall's τ, *Biometrical Journal*, **8**, 1003–9.

Scholz, F. W. (1978), Weighted median regression estimates, *The Annals of Statistics*, **6**, 603–9.

Schucany, W. R. (1978), Comments on a paper by M. Hollander and J. Sethuraman, *Biometrika*, **65**, 410–11.

Schucany, W. R. and J. Beckett, III (1976), Analysis of multiple sets of incomplete rankings, *Communications in Statistics – Theory and Methods*, **5**, 1327–34.

Schucany, W. R. and W. H. Frawley (1973), A rank test for two group concordance, *Psychometrika*, **38**, 249–58.

Schulman, R. S. (1979), A geometric model of rank correlation, *The American Statistician*, **33**, 77–80.

Schweizer, B. and E. F. Wolff (1981), On nonparametric measures of dependence for random variables, *The Annals of Statistics*, **9**, 879–85.

Sen, P. K. (1967), A note on the asymptotic efficiency of Friedman's χ_r^2 test, *Biometrika*, **54**, 677–9.

Sen, P. K. (1968), Estimates of the regression coefficient based on Kendall's tau, *Journal of the American Statistical Association*, **63**, 1379–89.

Sen, P. K. (1972), A further note on the asymptotic efficiency of Friedman's χ_r^2, *Metrika*, **18**, 234–7.

Sen, P. K. and P. R. Krishnaiah (1984), Selected tables for nonparametric statistics, in P. R. Krishnaiah and P. K. Sen (eds), *Nonparametric Statistics*, Elsevier Science Publishers, Amsterdam, pp. 937–58.

Shah, S. M. (1961), A note on Griffin's paper 'Graphic computation of tau as a coefficient of disarry', *Journal of the American Statistical Association*, **56**, 736.

Shirahata, S. (1977), Tests of partial correlation in a linear model, *Biometrika*, **64**, 162–4.

Shirahata, S. (1980), Rank tests of partial correlation, *Bulletin of Mathematical Statistics*, **19**(3–4), 9–18.

Shirahata, S. (1981), Intraclass rank tests for independence, *Biometrika*, **68**, 451–6.

Shirahata, S. (1985), Asymptotic properties of Kruskal–Wallis test and Friedman test in the analysis of variance models with random effects, *Communications in Statistics – Theory and Methods*, **14**, 1685–92.

Sievers, G. L. (1978), Weighted rank statistics for simple linear regression, *Journal of the American Statistical Association*, **73**, 628–31.

Sillitto, G. P. (1947), The distribution of Kendall's τ coefficient of rank correlation in rankings containing ties, *Biometrika*, **34**, 36–40.

Silva, C. and D. Quade (1983), Estimating the asymptotic relative efficiency of weighted rankings, *Communications in Statistics – Simulation and Computation*, **12**, 511–21.

Silverstone, H. (1950), A note on the cumulants of Kendall's S-distribution, *Biometrika*, **37**, 231–5.

Simon, G. (1977a), A nonparametric test of total independence based on Kendall's tau, *Biometrika*, **64**, 277–82.

Simon, G. (1977b), Multivariate generalization of Kendall's tau with application to data reduction, *Journal of the American Statistical Association*, **72**, 367–76.

Simon, G. A. (1978), Efficacies of measures of association for ordinal contingency tables, *Journal of the American Statistical Association*, **73**, 545–51.

Skillings, J. H. and G. A. Mack (1981), On the use of the Friedman-type statistic in balanced and unbalanced designs, *Technometrics*, **23**, 171–7.

Skillings, J. H. and D. A. Wolfe (1978), Distribution-free tests for ordered alternatives in a randomized block design, *Journal of the American Statistical Association*, **73**, 427–31.

Slater, P. (1961), Inconsistencies in a schedule of paired comparisons, *Biometrika*, **48**, 303–12.

Smid, L. J. (1956), On the distribution of the test statistic of Kendall and Wilcoxon when ties are present, *Statistica Neerlandica*, **10**, 205–14.

Smith, B. B. (1950), Discussion of Professor Ross's paper, *Journal of the Royal Statistical Society*, B, **12**, 54–6.

Snell, M. C. (1983), Recent literature on testing for intergroup concordance, *Applied Statistics*, **32**, 134–40.

Snow, B. A. S. (1962), The third moment of Kendall's tau in normal samples, *Biometrika*, **49**, 177–84.

Snow, B. A. S. (1963), The distribution of Kendall's tau for samples of four from a bivariate normal population with correlation ρ, *Biometrika*, **50**, 538–9.

Somers, R. H. (1959), The rank analogue of product-moment partial correlation and regression, with application to manifold, ordered contingency tables, *Biometrika*, **46**, 241–6.

Somers, R. H. (1962), A new asymptotic measure of association for ordinal variables, *American Sociological Review*, **27**, 799–811.

Somers, R. H. (1968), An approach to the multivariate analyses of ordinal data, *American Sociological Review*, **33**, 971–7.

Somers, R. H. (1974), Analysis of partial rank correlation measures based on the product-moment model: Part one, *Social Forces*, **53**, 229–46.

Spearman, C. (1904), The proof and measurement of association between two things, *American Journal of Psychology*, **15**, 72–101.

Spearman, C. (1906), A footrule for measuring correlation, *British Journal of Psychology*, **2**, 89–108.

Spearman, C. (1910), Correlation calculated from faulty data, *British Journal of Psychology*, **3**, 271–95.

Spurrier, J. D. and J. E. Hewett (1980), Two-stage test of independence using Kendall's statistic, *Biometrics*, **36**, 517–22.

Starks, T. H. and H. A. David (1961), Significance tests for paired-comparison experiments, *Biometrika*, **48**, 95–108.

Stuart, A. (1951), An application of the distribution of the ranking concordance coefficient, *Biometrika*, **38**, 33–42.

Stuart, A. (1953), The estimation and comparison of strengths of association in contingency tables, *Biometrika*, **40**, 105–10.

Stuart, A. (1954a), The correlation between variate-values and ranks in samples from a continuous distribution, *British Journal of Psychology*, (Statistical Section), **7**, 37–44.

Stuart, A. (1954b), Asymptotic relative efficiencies of distribution-free tests of randomness against normal alternatives, *Journal of the American Statistical Association*, **49**, 147–57.

Stuart, A. (1954c), The asymptotic relative efficiencies of tests and the derivatives of their power functions, *Skandinavisk Aktuarietidskrift*, **37**, 163–9.

Stuart, A. (1956a), Bounds for the variance of Kendall's rank correlation statistic, *Biometrika*, **43**, 474–7.

Stuart, A. (1956b), The efficiencies of tests for randomness against normal regression, *Journal of the American Statistical Association*, **51**, 285–7.

Stuart, A. (1963), Calculation of Spearman's rho for ordered two-way classifications, *The American Statistician*, **17**(4), 23–4.

'Student' (1921), An experimental determination of the probable error of Dr. Spearman's correlation coefficient, *Biometrika*, **13**, 263–82.

Sundrum, R. M. (1953a), A method of systematic sampling based on order properties, *Biometrika*, **40**, 452–6.

Sundrum, R. M. (1953b), Moments of the rank correlation coefficient τ in the general case, *Biometrika*, **40**, 409–20.

Sundrum, R. M. (1953c), The power of Wilcoxon's two-sample test, *Journal of the Royal Statistical Society, B*, **15**, 246–52.

Sundrum, R. M. (1953d), Theory and applications of distribution-free methods in statistics, *Ph.D. Thesis*, London University.

Taylor, W. L. (1964), Correcting the average rank correlation coefficient for ties in rankings, *Journal of the American Statistical Association*, **59**, 872–6.

Taylor, W. L. and C. Fong (1963), Some contributions to average rank correlation and to the distribution of the average rank correlation coefficient, *Journal of the American Statistical Association*, **58**, 756–69.

Teegarden, K. L. (1960), Critical $\sum d^2$ values for rank order correlations, *Industrial Quality Control*, **16**(11), 48–9.

Terpstra, T. J. (1952), The asymptotic normality and consistency of Kendall's test against trend when ties are present in one ranking, *Proceedings,*

Koninklijke Nederlandse Akademie van Wetenschappen, A, **55**, 327–33, *Indagationes Mathematicae*, **14**.

Terry, M. E. (1952), Some rank order tests which are most powerful against specific alternatives, *The Annals of Mathematical Statistics*, **23**, 346–66.

Theil, H. (1950), A rank invariant method of linear and polynomial regression analysis, I, II, III, *Proceedings, Koninklijke Nederlandse Akademie van Wetenschappen*, **53**, 386–92, 521–5, 1397–412; *Indagationes Mathematicae*, **12**, 85–91, 173–7, 467–82.

Thompson, G. L. and L. P. Ammann (1988), Efficacies of rank-transform statistics in two-way models with no interaction, *Journal of the American Statistical Association*, **84**, 325–30.

Tritchler, D. L. (1988), The exact nonparametric analysis of a class of experimental designs, *Communications in Statistics – Theory and Methods*, **17**, 1351–63.

Ury, H. K. and D. C. Kleinecke (1979), Tables of the distribution of Spearman's footrule, *Applied Statistics*, **28**, 271–5.

Van Dantzig, D. (1951), On the consistency and the power of Wilcoxon's two-sample test, *Proceedings, Koninklijke Nederlandse Akademie van Wetenschappen, A*, **54**, 1–8, *Indagationes Mathematicae*, **13**.

Van Dantzig, D. and J. Hemelrijk (1954), Statistical methods based on few assumptions, *Bulletin of the International Statistical Institute*, **34**, 2me livraison, 239–67.

Van der Laan, P. (1988), The use of Durbin's rank test, *The American Statistician*, **42**, 165.

Van der Laan, P. and J. Prakken (1972), Exact distribution of Durbin's distribution-free test statistic for balanced incomplete block designs, and comparison with the chi-square and *F* approximation, *Statistica Neerlandica*, **26**, 155–64.

Van der Vaart, H. R. (1950), Some remarks on the power function of Wilcoxon's test for the problem of two samples, *Proceedings, Koninklijke Nederlandse Akademie van Wetenschappen, A*, **53**, 494–520, *Indagationes Mathematicae*, **12**, 146–72.

Van der Waerden, B. L. (1952), Order test for the two-sample problem and their power, *Proceedings, Koninklijke Nederlandse Akademie van Wetenschappen, A*, **55**, 453–8; Corrigenda, *Ibid.* (1953), **56**, 80.

Van Elteren, P. and G. E. Noether (1959), The asymptotic efficiency of the χ_r^2 test for a balanced incomplete block design, *Biometrika*, **46**, 475–7.

Wallis, W. A. (1939), The correlation ratio for ranked data, *Journal of the American Statistical Association*, **34**, 139–51.

Watkins, G. P. (1933), An ordinal index of correlation, *Journal of the American Statistical Association*, **28**, 139–51.

Wei, L. J. (1982), Asymptotically distribution-free simultaneous confidence regions for treatment differences in a randomized complete block design, *Journal of the Royal Statistical Society, B*, **44**, 201–8.

Wei, T. H. (1952), The algebraic foundations of ranking theory, *Ph.D. Thesis*, Cambridge University.

Weier, D. R. and A. P. Basu (1978), A trivariate generalization of Spearman's rho, *Technical Report No. 78*, Department of Statistics, University of Missouri, Columbia, MO.

Welch, B. L. (1937), On the z-test in randomised blocks and Latin Squares, *Biometrika*, **29**, 21–52.

Whitfield, J. W. (1947), Rank correlation between two variables, one of which is ranked, the other dichotomous, *Biometrika*, **34**, 292–6.

Whitfield, J. W. (1949), Intra-class rank correlation, *Biometrika*, **36**, 463–7.

Whitfield, J. W. (1950), Uses of ranking method in psychology, *Journal of the Royal Statistical Society, B*, **12**, 163–70.

Wilcoxon, F. (1945), Individual comparisons by ranking methods, *Biometrics*, **1**, 80–3.

Wilkie, D. (1980), Pictorial representation of Kendall's rank correlation coefficient, *Teaching Statistics*, **2**, 76–8.

Wilkinson, J. W. (1957), An analysis of paired comparisons designs with incomplete repetitions, *Biometrika*, **44**, 97–113.

Wolfe, D. A. (1977), A distribution-free test for related correlation coefficients, *Technometrics*, **19**, 507–9.

Wood, J. T. (1970), A variance stabilizing transformation for coefficients of concordance and for Spearman's rho and Kendall's tau, *Biometrika*, **57**, 619–27.

Woodbury, M. A. (1940), Rank correlation when there are equal variates, *The Annals of Mathematical Statistics*, **11**, 358–62.

Woodworth, G. G. (1970), Large deviations and Bahadur efficiency of linear rank statistics, *The Annals of Mathematical Statistics*, **41**, 251–83.

Yohai, V. J. and N. E. Ferretti (1987), Tests based on weighted rankings in complete blocks: Exact distribution and Monte Carlo simulation, *Communications in Statistics – Simulation and Computation*, **16**, 333–47.

Zar, J. H. (1972), Significance testing of the Spearman rank correlation coefficient, *Journal of the American Statistical Association*, **67**, 578–80.

References for applications

Ashton, A. H. (1985), Does consensus imply accuracy in accounting studies of decision making? *The Accounting Review*, **55**(2), 173–85.

Bonjean, C. M., B. J. Brown, B. D. Grandjean and P. O. Macken (1982), Increasing work satisfaction through organizational change: A longitudinal study of nursing educators, *Journal of Applied Behavioral Science*, **18**(3), 357–69.

Cancian, F. M. and B. L. Ross (1981), Mass media and the women's movement: 1900–1977, *Journal of Applied Behavioral Science*, **17**, 9–26.

Charlop, M. H. and J. Carlson (1983), Reversal and nonreversal shifts in autistic children, *Journal of Experimental Child Psychology*, **36**, 56–67.

Davis, H. L. (1970), Dimensions of marital roles in consumer decision making, *Journal of Marketing Research*, **7**(May), 168–77.

Dickson, P. R., R. F. Lusch and W. Wilkie (1983), Consumer acquisition priorities for home appliances: A replication and re-evaluation, *Journal of Consumer Research*, **9**(4), 432–5.

Dominguez, L. and C. Vanmarcke (1987), Market structure and marketing behavior in LDCs: The case of Venezuela, *Journal of Macromarketing*, **7**(2), 4–16.

Dubinsky, A. J. and W. Rudelius (1980–81), Selling techniques for industrial products and services: Are they different? *Journal of Personal Selling and Sales Management*, **1**(1), 65–74.

Duckson, Jr., D. W. (1988), Patterns of spatial awareness, *Journal of Research and Development in Education*, **21**(2), 30–5.

Durbin, J. (1951), Incomplete blocks in ranking experiments, *British Journal of Psychology* (Statistical Section), **4**, 85–90.

Elbert, D. J. and D. G. Anderson (1984), Student reaction to local vs. published cases: Is there a difference? *Journal of Marketing Education*, **6**(Summer), 18–21.

Ezell, H. F. and P. J. Ward (1983), Differentiating between frequent and less frequent patrons of beauty salons: A focus on working women, Southern Marketing Association, *Proceedings 1983*, 68–71.

Fienberg, S. E. (1971), Randomization and social affairs: The 1970 draft lottery, *Science*, **171**(Jan.), 255–61.

Goss, F. L. and C. Karam (1987), The effects of glycogen supercompensation on the electrocardiographic response during exercise, *Research Quarterly for Exercise and Sport*, **58**(1), 68–71.

Hafer, J. C. and C. C. Hoth (1981), Grooming your marketing students to match the employer's ideal job candidate, *Journal of Marketing Education*, **3**(Spring), 15–19.

Hartley, L. R. (1973), Effect of prior noise or prior performance on serial reaction, *Journal of Experimental Psychology*, **101**(2), 255–61.

Holden, M. and P. Holden (1978), Effective tariff protection and resource allocation: A nonparametric approach, *Review of Economics and Statistics*, **60**, 294–300.

Hopwood, W. and J. McKeown (1987), Evidence on surrogates for earnings expectations within a capital market context, unpublished paper presented at Accounting Workshop, School of Accountancy, Tuscaloosa, AL.

Howell, W. C. and L. T. Johnson (1982), An evaluation of the compressed-course format for instruction in accounting, *The Accounting Review*, **57**(2), 403–13.

Humphreys, A. and P. K. Smith (1987), Rough and tumble, friendship and dominance in school children: Evidence for continuity and change with age, *Child Development*, **58**, 201–12.

Ingram, R. W. (1984), Economic incentives and the choice of state government accounting practices, *Journal of Accounting Research*, **22**(1), 126–44.

Jackman, M. K. (1985), The recognition of tachistopically presented words, varying in imagery, part of speech and word frequency, in left and right visual fields, *British Journal of Psychology*, **76**, 59–74.

Juchau, R. and M. Galvin (1984), Communication skills of accountants in Australia, *Accounting and Finance*, **24**(1), 17–32.

Lawrence, J. and D. M. Steed (1984), European voices on disruptive behavior in schools, *British Journal of Educational Studies*, **32**(1), 4–17.

Lawrence, M. J. (1983), An exploration of some practical issues in the use of quantitative forecasting models, *Journal of Forecasting*, **2**(2), 169–79.

Lusch, R. F. and R. H. Ross (1985), The nature of power in a marketing channel, *Journal of the Academy of Marketing Science*, **13**(3), 39–56.

McCaslin, T. E. and K. G. Stanga (1986), Similarities in measurement needs of equity investors and creditors, *Accounting and Business Research*, **16**(Spring), 151–6.

McDaniel, S. W. and R. T. Hise (1984), Shaping the marketing curriculum: The CEO perspective, *Journal of Marketing Education*, **6**(Summer), 27–32.

Maghsoodloo, S. (1975), Estimates of the quantiles of Kendall's partial rank correlation coefficient and additional quantile estimates, *Journal of Statistical Computation and Simulation*, **4**, 155–64.

Mitman, A. L., J. R. Mergandoller, V. A. Marchman and M. J. Packer (1987), Scientific literacy, *American Educational Research Journal*, **24**, 611–33.

Muczyk, J. P. and M. Gable (1981), Unidimensional (global) vs. multidimensional composite performance appraisals of store managers, *Journal of the Academy of Marketing Science*, **9**(3), 191–205.

Peters, W. S. (1987), *Counting for Something*, Springer-Verlag, New York, p. 146.

Reierson, C. (1966), Are foreign products seen as national stereotypes? *Journal of Retailing*, **42**(3), 33–40.

Rickey, B. (1954), Goodbye to some old baseball ideas, *Life Magazine*, August 2, 79–89.

Robichaud, M. and W. Wilson (1976), Rank order of preference of blacks and whites, *Southern Journal of Education Research*, **10**, 156–66.

Rosenberg, L. J., C. K. Gibson and D. B. Epley (1981), How to retain real estate sales people, *Journal of Personal Selling and Sales Management*, 1(2), 36–42.

Russell, R. S. and B. W. Taylor III (1985), An evaluation of sequencing rules for an assembly shop, *Decision Science*, 16, 196–212.

Sands, R. G., L. G. Newby and R. A. Greenberg (1981), Labeling of health risk in industrial settings, *Journal of Applied Behavioral Science*, 17, 359–74.

Shearer, L. (1982), Pierpont's presidential report card, *Parade Magazine*, January 7, 1982.

Sheehan, J. M., T. W. Rowland and E. J. Burke (1987), A comparison of four treadmill protocols for determination of maximum oxygen uptake in 10 to 12 year old boys, *International Journal of Sports Medicine*, 8, 31–4.

Shipley, D. D. (1984), Selection and motivation of distribution intermediaries, *Industrial Marketing Management*, 13, 249–56.

Sinclair, E., D. Guthrie and S. R. Forness (1984), Establishing a connection between severity of learning disabilities and classroom attention problems, *Journal of Educational Research*, 78(1), 18–21.

Slovic, P., B. Fischhoff and S. Lichtenstein (1980), Risky assumptions, *Psychology Today*, June, 45–7.

Stuart, A. (1953), The estimation and comparison of strengths of association in contingency tables, *Biometrika*, 40, 105–10.

Whipple, T. W. and L. A. Neidell (1971–1972), Black and white perceptions of competing stores, *Journal of Retailing*, 47(4), 5–20.

Wind, Y., V. Mahajan and D. J. Swire (1983), An empirical comparison of standardized portfolio models, *Journal of Marketing*, 47(Spring), 88–89.

Index

Agreement, coefficient of 188–93, 198–200
Analysis of variance by ranks (*see* Friedman test)
Association
  bivariate population 164–5
  contingency table 50–5
  m sample measure of
    coefficient of concordance 118
    complete rankings 117–27
    Goodman–Kruskal coefficient 52–4
    incomplete rankings 129–33
  two sample measure of
    Kendall's tau 3–8, 10–20
    point-biserial 47–9
    rho 8–20
    Spearman's rho 8–20
    tau 3–8, 10–20
Asymptotic relative efficiency 115–16, 134, 203

Circular triads 185–7, 195–8
Coefficient of concordance (*see* Concordance, coefficient of)
Compact set of ranks 34
Concordance, bivariate population
  definition 164–5
  estimators of 176–83
  probability of 164–5
Concordance, coefficient of
  complete rankings 117–27
  definition 118
  incomplete rankings 129–33, 151–3
  probability distribution 150–3
  relation with rho 119
  test of independence 121–4, 144–8
  table references 133
  ties correction 124
Concordance, two-group 201–2
Concordance coefficient (*see* Tau)
Concordance of pairs of ranks
  definition 6–7
  parameter estimation 176–7
  probability of 164–5
  types 33
Confidence interval for tau 74–6, 110–13
Conjugate rankings 10–11
Consistence, coefficient of 186–8

Contingency tables 49–55
Continuity correction
  coefficient of concordance 123–4
  rho 70–1, 103
  tau 65–9, 103
Correction for ties
  coefficient of concordance 124
  paired comparisons 190–1
  rho 43–7
  tau 40–3, 47–54, 66–67
Correlation (*see* also Association)
  between ranks and variate values 164–83
  generalised 25
  grade 169–70
  partial rank 154–61
  product-moment 167
  rank 3
Criterion ranking comparisons 202

Daniels inequality 12, 20, 34
Daniels test for trend 76–7
Dichotomy, as a ranking 49–50
Disarray
  rho as a coefficient of 9–10, 31–3
  tau as a coefficient of 7–8, 29–31
Discordance 6–7, 33
Durbin test 129–35, 151–3
Durbin–Stuart inequality 12–13, 20, 34–7

Efficiency of
  coefficient of concordance 134
  Friedman test 134
  regression 203
  rho 115–16
  tau 115–6
Estimation of
  population consensus 124–7, 201
  product-moment correlation 167–8
  rho 165
  tau 164–5
  true ranking 124–7, 151

Fisher's *z* distribution 122–3, 131–3, 144–9
Friedman test
  definition 127–8, 133
  efficiency 134
  table references 133

General rank correlation   25–39
Goodman–Kruskal coefficient   52–4
Grade correlation   169–70

Incomplete Latin square   129
Incomplete rankings   129–33, 134–5, 151–3
Independence, tests of
    concordance coefficient (complete
        rankings)   121–4, 144–51
    concordance coefficient (incomplete
        rankings)   129–33, 151–3
    rank correlation   60–79
    tau   60–79
Interchanges (*see* Disarray)
Inversions   9–10, 20, 31–33

Kendall coefficient of concordance (*see*
    Concordance, coefficient of)
Kendall tau (*see* Tau)

Likert scale data   18–19

m-rankings, problem of   117–53
Mann test for trend   76–7
Midranks   40

Order properties   164–5
Ordered contingency tables   49–55

Paired comparisons   184–200
Partial rank correlation   154–61
Pearson product-moment correlation
    coefficient   27, 167
Population of rankers   201
Preference (*see* Paired comparisons)
Product-moment correlation coefficient
    between rho and tau   102
    between variate values and ranks   165–6,
        173–6
    definition   27
    special case of general rank correlation   27
P-value   64–5

Rank, definition   1
Rank correlation coefficient (*see* Rho, Tau)
Regression   202–3
Rho (r_s)
    coefficient of disarray   9–10, 31–3
    continuity correction   70–1, 103
    definition   8–9
    efficiency   115–16
    independence, use in testing   60–79

joint distribution with tau   99–102
non-null distribution   113–16
null distribution   69–71, 97–9
population parameter analogy   165
relation with concordance coefficient   119
relation with product-moment
    correlation   169–71
relation with tau   171
special case of general rank
    correlation   26–7
table references   78
test of independence   60–79
ties correction   43–7
trend, use in testing   76–9

Spearman rank correlation (*see* Rho)
Spearman's footrule   37–9
Stragglers   13–15

Tau (t)
    calculation   5–6
    coefficient of concordance   6–7
    coefficient of disarray   7–8, 29–31
    continuity correction   65–9, 103
    definition   3–5
    efficiency   115–16
    independence, use in testing   60–79
    joint distribution with rho   99–102
    non-null distribution   72–6, 104–13,
        115–16
    null distribution   60–5, 90–7
    partial   154–61
    population parameter analogy   62, 164–5
    relation with product-moment
        correlation   166–9
    relation with rho   171
    special case of general rank
        correlation   25–6
    table references   78
    test of independence   60–79
    ties correction   40–3, 47–54, 190–1
    trend, use in testing   76–9
Tau test for independence   60–79
Tied ranks   40–55, 66–7, 124, 190–1
Time series data   13–15
Trend, tests for   76–79
Triads, in paired comparisons   185–7
Two-group concordance   201–2

U statistics   38

Youden square   129